AF887962

RURAL ELECTRIC AND SUPPLY COOPERATIVES WERE MY CONCERN

AN AUTOBIOGRAPHY BY
WALTER G. SCHMIDT
CERTIFIED PUBLIC ACCOUNTANT

WL-PAN PRESS
Kansas City, Missouri 1987

FIRST PRINTING 1987

Library of Congress Catalog Card Number 86-91678

Schmidt, Walter G. CPA
Rural Electric And Supply Cooperatives
Were My Concern

1. Publishers and publishing—Biographies etc.

ISBN 0-940533-00-6

PRINTED IN THE UNITED STATES

CONTENTS

ACKNOWLEDGEMENTS

I am indebted to my brothers and sisters, Henry, Regina, Marie, Ray, Elizabeth and Gilbert for their assistance in developing the history of life on the farm.

My gratitude goes to my wife Louise and my daughters Paula, Anita and Nancy for their assistance in composition of the book.

DEDICATION

I want to dedicate this book to all the people involved in rural cooperatives. To the local members and the members of the board of directors who saw fit to employ me as their manager. To the management, employees and members of Farmland Industries, Inc. that I was privileged to serve. To the Rural Electrification Administration, the REA Borrower clients, their directors and employees who made life pleasant for me for thirty-two years.

FOREWORD

Twenty-six percent of this book deals with my ancestors' migrations to this country, my parents' wheat farming, our farm family and the struggle to obtain a formal education. Eighteen percent of the book covers about twenty years of my life as an employee of rural-supply cooperatives both local and regional including manufacturing, wholesaling and retailing. Thirty-five percent of the book is devoted to my public accounting practice as a certified public accountant spanning over thirty-two years and specializing in the auditing of REA Borrowers. For eleven of those years the audits were conducted as a private practitioner, while I practiced as a senior partner of Schmidt & Company, CPAs the last twenty-one years. Seventeen percent of the book takes you with us on our overseas travels and four percent of the book discloses some of my philosophies.

The first part of the book should be of interest to many people living in Rush and Ellis County, Kansas. My parents are buried in St. Joseph's Cemetery in Hays, Kansas. Those counties of Kansas are primarily agricultural areas and no doubt many remember the dust storms and the terrible economic depression that hit farmers in the thirties. Some of the names mentioned in the book will be recognized. One of the reasons given for buying land along the Smoky River and not the Pawnee River was that land in Pawnee County cost four dollars per acre and land along the Smoky River cost two and one-half dollars per acre. I think anyone familiar with land in those areas would agree they should have purchased the land in Pawnee County. Some of the less expensive land proved to be an excellent purchase when oil was discovered underneath it. We were not that fortunate.

The book should be interesting to my seventy nephews and nieces scattered over the country. All of them will learn they came from a long line of Germans who became farmers not by choice but by necessity. They will also learn that their Roman catholic faith has followed those Germans for over two hundred years. The book's description of living and working on the farm will be of interest to the older ones. Some of them were born and worked on a farm. They will recognize the machines and tools used in farming in the thirties. Some will remember how hard one had to work to earn a dollar.

Hopefully the book of my life will convince my younger relatives and all young people for that matter, that one does not have to be born with a silver spoon in one's mouth to succeed in this great free country. Success, however, must be earned. There are no free lunches. You can be master of your own destiny if you are willing to pay the price. If your destiny in life is to improve your standard of living then you must set a goal and work hard to reach that goal. Remember, "If at first you don't succeed, try, try again." Persistence is a very important ingredient. We live in a competitive society and you must train yourself to stand erect under pressure. As Harry Truman said, "If you can't take the heat, get out of the kitchen." One should not become despondent with the first stumbling block. Life has many hurdles and after conquering the first one, the others will not appear so insurmountable. Work is part of the answer but unless you become a likeable person who elicits respect and goodwill from your peers, your employer and the public, all the effort applied may be futile. What we think of ourselves is very important, but what others think of us is crucial. Show me a person who says, "I don't give a damn what others think" and I will show you a loser.

Managers and directors of local rural cooperatives will find my experience as a local manager amazingly similar to the problems encountered today. One of my nephews is a director of a local cooperative and when I told him, "Cooperatives are making the same mistakes made fifty years ago. It is my understanding that many cooperatives have their books full of accounts receivable and no cash in the bank." His response was, "Uncle Walt, you will just have to let us make our own mistakes." Unlike the Chinese, we do not respect prior generations. It appears that each generation will only learn from its own mistakes.

Names used in this part of the book will be familiar in Rush County. Some of the experiences I disclose in the book have never been a part of the annual report or disclosed by the local newspaper. I had been exposed to the Farmers Union organization but the concept of consumers cooperatives was rather new and challenging. The resistance to cooperatives by competing enterprises was short sighted and ill advised. Both rural-supply cooperatives and rural-electric cooperatives were accused of being socialistic. The free enterprise system is not restricted to a few existing enterprises. The job at LaCrosse, Kansas, gave me some very necessary management and public relations experience. My record at LaCrosse caught the attention of others and eventually brought me to Kansas City. Hundreds of people in Rush County recognize my name to this day.

As stated in the introductory paragraph of the Foreword, my experience with Farmland Industries, Inc., (CCA) was always in the area of management and auditing. The attraction that accounting had for me in high school remained with me as time passed. My first assignment was resurrecting local cooperatives in financial trouble (Managed Stations) and to conduct annual audits for other local cooperatives. This is a true story of what I saw and heard inside Farmland Industries, Inc. and portions of the story have never been published previously. Since this organization is the largest farm supply cooperative in the United States it brought me in personal touch with thousands of cooperative employees and directors.

Many should find this part of the book interesting reading. The first five years of my experience was devoted to problems connected with local cooperative members. The last five years concerned the activities of the manufacturing and control divisions of Farmland (CCA).

The largest part of the book and for the major portion of my adult life, I have practiced as a Certified Public Accountant. Eleven years as a private practitioner and for twenty-one years as a senior partner of Schmidt & Company, CPAs. I acquired my CPA Certificate in 1951 while I was employed by Farmland (CCA). The Rural Electrification Program was about fifteen years old then and virtually all of the borrowers were rural cooperatives. The CPA audit of these borrowers became available in 1952 and 1953 in the Midwest and since I was known in the rural areas, many of the borrowers became audit clients of mine. When I retired in 1984, Schmidt & Company, CPA's was conducting more electric REA Borrower audits than any CPA firm in the country.

This part of the book covering my thirty-two years auditing REA Borrowers will create a good deal of interest in the Midwest where our audits were conducted and even more interest across the land where about a thousand rural electric borrowers and about the same number of rural telephone borrowers engage a CPA firm to conduct their annual audit. It will be interesting reading by the CPA profession and the professional societies since about five hundred firms are involved in the audits of REA Borrowers. My experience with the program is a story of contrasts: A story of unbelievable support from clients, CPA professionals and REA field personnel while some REA personnel in Washington, D.C. conducted a campaign of denigration and harassment unparalleled. It is a true story that has never been told. Since so many of our clients became involved, it will bring back memories of meetings at Atlantic City when the flood came waltzing in. Rural electrification is one of the finest programs ever undertaken by our government. Without it a large part of Rural America would still be in the dark. The administration of the program should come under review and evaluation but the withdrawal of the government at this time would kill the very project undertaken in the first place. The REA loan program is fifty years old. The only question raised in this book might be, "Has the REA Administration encouraged higher members' equity and local ownership?"

The true story of Schmidt & Company, CPAs, a partnership, its origin, its policies and its problems, are enumerated in this part of the book. Provisions for office management, number of partners, wage policy, expense reimbursement and retirement pension should be of interest to private practitioners anticipating forming a partnership. I completed my retirement from the partnership at the end of 1984 and Schmidt & Company, CPA's continues to render services to REA Borrowers under the control of successor senior partners.

Anyone anticipating travel overseas might find the portion of the book reporting on our travels to our fiftieth state, as well as, our travels to West Germany, Italy, Austria, Czechoslovakia, Switzerland, England, Wales, Scotland and Ireland interesting.

The last chapter of the book enumerates the author's opinion regarding such subjects as: Importance of character, religious belief, family, success or failure, the economy, political parties, government versus private enterprise, social security system, PAC committees, liability insurance problems and a life style that leads to contentment and peace.

KANSAS CITY, MISSOURI — Walter G. Schmidt, CPA

CHAPTER I

ROOTS IN GERMANY AND RUSSIA

When my ancestors arrived in this country one hundred and ten years ago they were referred to as "German Russians" or as "Volga Germans" and this chapter explains the migrations that account for those expressions. Probably the only Russian that could persuade my ancestors to move to Russia was Catherine the Great, Empress of Russia. They trusted her because she was a German by birth. Germans didn't like the Russians two hundred years ago and they don't like them today.

SOME people go to great lengths to discover their roots, so let me introduce you to mine. My great, great grandfathers on both the paternal and maternal side of the family were born in southern Germany in what is now Bavaria. Available information indicates that neither of them was a farmer in Germany. The Schmidts appear to have been affiliated with the leather industry in Germany and the Millers (Muellers) appear to have been associated with the milling industry.

About the time the American Colonies and England severed their relationship with the Declaration of Independence on July 4, 1776, my ancestors were so disenchanted with the economic conditions of their country, the continuing wars among the principalities of Germany and the decades of conflict among the provinces between those allied with Martin Luther and those remaining loyal to the Roman Catholic Church, any offer of relief was welcome. Germany was devastated by Seven Year's of war between Prussia and Austria (1756-1763). In the meantime Great Britain and France were struggling for supremacy in North America and on the sea. The German peasants were very poor and about 30,000 volunteered as mercenaries to help England in the American Revolution.

As the movie "Peter the Great" depicts, rumblings coming out of Russia indicated the Czars were determined to move their country from its isolationist policy to one of communication and trade with the nations of the world. Access to sea routes through the Baltic and the Black sea were acquired and trade agreements were signed with England and others. With the death of Peter III, Catherine II, a German by birth, ascended to the throne as Empress of Russia in 1762 and ruled the empire until 1796. Catherine the Great, an energetic and astute ruler, continued to develop her empire in accordance with the policies of Peter the Great. One of her immediate goals was to import people from other nations to inhabit and develop the large tracts of land near the lower Volga River being invaded by nomadic tribes from Asia.

On December 4, 1762, Catherine the Great drafted a document called a "manifesto" and flooded Europe with literature and pamphlets, all of them offering special privileges to anyone who would settle the Volga Region. Six months passed and the response to the invitation was very disappointing.

On July 22, 1763, Catherine the Great issued a revised invitation or manifesto offering the immigrants the following:

1. Right of self-government.
2. Freedom of religion.
3. Right to build churches and schools. (No monasteries)
4. Freedom to pursue their old professions.
5. The right to have priests and teachers.
6. Free transportation and living expenses for the migration.
7. Loans, interest free, for ten years for homes, stock and tools.
8. No taxes, levies or land service for thirty years.
9. Exemption from military duty for an indefinite time. (Later interpreted as one hundred years.)

The sole requirement was that the immigrants be willing to take the oath of allegiance in accordance with their own religious rites and to submit to the prevailing form of civil law.

This time Catherine the Great made sure the message reached its destination, by sending her own deputies and rewarding them in direct proportion to the number of recruits delivered. Additional benefits were promised the people by the deputies such as the privilege of leaving Russia, providing a portion of the property acquired in Russia be left behind. The deputies were to accompany the recruits to the promised land, assist them in establishing colonies and help provide food and shelter. In assigning migrants to colonies, consideration was given to common language, religious belief etc.

Between 1763 and 1768, when Joseph II, the Emperor of the Holy Roman Empire, put an end to the migration, some 8,000 families, 27,000 people in all, migrated from Western Europe to Russia. Most of them came from Germany.

The itinerary for the migrants from Western Europe called for collection stations in various cities, transportation by wagon and on foot north to Lubek on the Baltic Sea (about 500 miles), by boat from Lubek to Leningrad (about 500 miles) and overland by wagon and on foot to Lomonov, the location of the Imperial Summer Palace, where Catherine met the migrants in person. Here the settlers took their oath of fealty to the Crown of Imperial Russia, in a church of their faith. The oath was read to them in German. From there they traveled by land and rivers to Tver (about 100 miles) and finally by boat on the Volga to Saratov. The Volga is Europe's largest river and extends over 2,293 miles. Traveling down the Volga might take five to six weeks if one camped out at night. The trip by land could be made in two to three weeks. For some, the trip to their destination took over a year.

Once people signed on for the trip, there was no turning back. Guards made certain of that. When one of the groups arrived at Lubek, low and behold, there were no Russian ships, instead wooden barracks had been erected as temporary housing for the winter. Catherine's representatives explained this was necessary because it was September and ice would be forming along the Russian rivers, as well as, the harbors along the seacoast. The barracks were very crowded, unclean and very cold during the winter. Quarreling and fighting took place. Some contracted diseases and died. The first contingent was allowed to sail April 1766. If the ships were not delayed by strong winds, the trip to Leningrad took about eleven days.

Some of the German passengers suspected that some of the ships were delayed for weeks so that more supplies could be sold. Bearded Russian teamsters greeted them at Leningrad and migrants discovered that promises of opportunities for business and professional people were a farce. Almost half of the migrants were artisans, having no knowledge of farming. Once in the Land of the Czars, however, all without exception were transported to the Lower Volga. The first settlers arrived at Saratov June 20, 1767.

To the astonishment of the settlers, the vaunted paradise of the deputies, proved to be a vast empty expanse of wild, semi-arid steppe land. Moreover, the buildings promised by the government had not been erected. The allowance of money advanced by Her Majesty proved sadly insufficient. To make matters worse, the colonists had arrived too late in the season to do any planting. The only shelters they would be able to construct were the Russian style dugouts. The dugout was a simple affair, a hole was excavated and the earth from the excavation banked along the perimeter. The roof consisted of branches or poles laid close together and sealed with a layer of earth or clay. The dugouts were dark, inclined to be damp and there was virtually no ventilation. It had one small virtue. Since the dugout was well below the frost line, it used a minimum of fuel in the winter. As many as four families might share one dugout.

The colonizing of the Volga cost the Crown 5,199,813 rubles and that was the amount charged to the settlers. Before my ancestors left Russia a compromise was reached on April 20, 1872 and the debt was reduced to 1,210,197 rubles. This amount represented funds used for building homes and churches, for medicine, to compensate the settlers for lives lost on the trip and losses due to raids by wild tribes. The government designated an area of about 133 miles on the east, or valley-side, of the Volga and about 92 miles on the west, or hilly-side, of the Volga for colonization. The entire region was divided into districts of about 50 square miles. Adult males could claim about 65 acres and all land would revert back to the government when it was divided for the next occupant, except the eleven acres on which the buildings were situated. The government also provided land for schools and churches. Although Catherine kept the colonies under supervision, she gave them the right to govern themselves. Each colony had a mayor, two committee men and a clerk. They were elected for two years and constituted the village board. Their decisions were law in the villages. The duties of the board included hiring a teacher, providing churches, collecting taxes and punishing misdemeanors.

The colonies also had a court system. Fifteen or sixteen villages selected delegates and these elected a district judge, two assistants, a secretary, a bookkeeper, a translator, a physician and a surveyor. These were elected for three years. The judge could hold court and hear cases from the fifteen or sixteen villages. However, the circuit court was subject to the jurisdiction of the governing body in Saratov. While the Russians gave each colony a name, the Germans usually named the colony after the first mayor and ignored the Russian name.

The villages were built in circles and were fortified with walls and ditches. A sentinel sat in the church tower. At night the settlers built fires to keep the wolf, lynx and bears at bay. During the day, work in the fields was done in companies in order to protect themselves and their stock from roving bands. The Kirgiz were a nomadic Tarter tribe that lived between the lower Volga and the Caspian Sea. They did not appreciate the intrusion of the German immigrants and sought to drive them out. They often stormed the villages, plundering, killing, burning, raping and dragging women and children into captivity. The villages were also attacked by a gang of peasants organized by a cossack to fight oppressive landlords. This revolt became so strong that the government finally had to suppress it with armed force. The cossack was captured and executed on January 10, 1775. The settlers appeal for government help brought little aid for years and it has been estimated that some 1,200 men, women and children were taken into captivity in those raids. A Russian army force of 250 men finally defeated the Kirgiz on August 16, 1774 and freed prisoners taken. Russian outposts were then established on the Ural border. After August 1774, no major raids were made in the Volga region.

A government agency provided the settlers with seed for wheat, rye and barley. Each family was given two broken-down nomad horses, a hand-made wooden plow, a sickle, hatchet, rake, wagon, a cow and a few necessary food supplies. The tailors, the artisans, the teachers, barbers and tradesmen were completely out of their elements. Some of them made the necessary adjustment and became successful farmers while others were never able to adjust to the situation. The first good harvest was in 1776. Wheat was cut with scythes and sickles, the women followed with rakes and twine, gathering the wheat and tying it into sheaves. The sheaves were put into shocks where they were left to dry. The shocks were then hauled to the village where the kernels were separated from the straw. The wheat was put on a large plot and horses pulling stone rollers were driven again and again over the wheat thus separating the kernels from the chaff. The grain was further separated from the straw by running it through a hand-powered mill or by allowing the wind to separate the grain from the straw. The grain was sacked and stored in a granary. Farm products were marketed by organizing a caravan of wagons for safety sake and delivering them to Saratov, a two-day trip. Farming was done at a distance of ten or more miles from the village at times. When the distance became too great to return to the village each evening, provisions such as food, water and tents were taken to the fields and farming was accomplished for several days before returning to the village.

The fathers and mothers of the household were automatically respected as the heads of the household, while the oldest son and his wife controlled the younger married couples and their families. The "Patriarchal System" was brought from Germany to Russia and later it was the custom in the United States. As second male generations grew to manhood and married, they lived with their parents because land was scarce and there was no money with which to set up homes on their own. Furthermore, they realized that it was more profitable to work together. The settlers clung tenaciously to their Roman Catholic Religion and their German language. Education for their children was slow in coming and that was especially true for girls of the family. The German people steadfastly refused to learn Russian or have their children taught the Russian language. Due to the friendly relations that existed between the two denominations, the Catholics and Lutherans compromised and helped each other. The priests baptized, performed marriage ceremonies and conducted funerals for the Lutherans on the hilly-side while the Lutheran ministers reciprocated for the villages on the valley-side. Despite all hardships, religion came first in their lives. The greater the suffering the more they clung to their faith. At times it was their only hope.

With freedom from marauding tribes secured, the colonists turned their attention and efforts toward a better life for themselves. Family homes consisted of a courtyard surrounded by a hedge with an opening or gate. Buildings inside the courtyard included a square house, behind the house a summer kitchen and behind that a wooden barrier, the barn, granary and sheds. Apparently the courtyard was large enough to provide some fruit trees and a vegetable garden. The barn was not large enough to house many animals. The windows of the house had shutters that were closed to conserve heat in the winter and to provide protection from invaders. The summer kitchen was used as a workshop for washing, summer cooking, smoking meat and for sausage making. The garden provided vegetables while animals were raised for the meat requirements of their diet. Temperatures were low enough so additional refrigeration was not required for a good part of the year. As years passed, the first houses were replaced by better buildings, the wooden hook plow was replaced by the one-share plow, the sickle by the cradle and the Russian pole wagon by the German rack or shaft wagon. Windmills were built. The crops were wheat, barley, oats, sugar beets, tobacco, potatoes and sunflower seeds. As the population grew and land was limited to the original amount given to the settlers, it became more and more difficult to make ends meet. The land area per individual had been reduced in some instances to one-fifth of the original. Many of the villages had three to four thousand people and farms often lay long distances from the village. Starting about 1845 additional grants of land were made by the government resulting in many new colonies.

My German ancestors lived in Russia for 109 years (from the summer of 1767 to the summer of 1876) and for the first half of that century, through hard work and fervent prayer, they extracted from the land a standard of living that became the envy of the Russian peasant. Part of their success was undoubtedly attributable to exemption from taxes and military duties granted to them by Empress Catherine the Great. The Russian peasant resented what he regarded as preferential treatment accorded the Germans. With the death of Catherine II in 1796 the good relationship between the colonists and the Czars, in existance for about thirty years, suffered gradual deterioration. The Germans loved the clear blue waters of the Volga, the wild flowers growing along its banks, the beautiful new churches erected in the villages and their associations with neighbors and relatives. Nevertheless, they discerned that public opinion was exerting pressure on the Czar to assimilate the Germans into the Russian society. The Czars and the Germans coexisted in this precarious arrangement for about twenty-five years but starting about 1850 the relationship with the government changed.

The government acceded to pressure from the people starting with 1850 and passed a number of laws changing the favored status of the colonists. On June 4, 1871, the Codex of the colonies, that contained the special rules applicable to the German colonies, was abolished. The Volga Germans did not perceive themselves as Russians. They did not like their Russian neighbors nor their way of life and chose to remain by themselves in the German colonies. These people believed that the Czar had deceived them. Now they began to grasp the importance of the Oath of Fealty to the Czar they made to Catherine II over a hundred years before.

In June, 1871, an edict issued by Alexander II limited the period of exemption from military service for the colonists to ten years but it did permit the colonists to emigrate within that time without forfeiture of property. Chancellor Bismarck of Germany used his influence to obtain this privilege for the colonists. November 24, 1874 four boys from Herzog and twenty-one boys from Katarinenstadt were drafted. It is estimated there were about 260,000 Germans in the Volga region when my ancestors started their migration to the United States. The migration out of Russia appeared to include a relatively small portion of the German population of the Volga region at that time. Before leaving, each one had to obtain a passport from the circuit judge and these were not issued to criminals or to individuals owing money. A release from the village had to be approved by two-thirds of the heads of families. □

CHAPTER II

MIGRATION TO THE NEW WORLD

Their faith in Catherine the Great began to wane when they discovered upon arrival in Russia that her deputies grossly misrepresented what was awaiting them. The Empress kept her word however with respect to land grants and self government. My ancestors were relatively prosperous and happy under the rule of Catherine the Great. After her death, the Russian influence and envy turned against the Germans. My ancestors left family and friends behind and migrated to the United States. This chapter describes their migration.

IN VIEW of all the problems encountered by the migration of their parents and grandparents from Germany to Russia and the years of toil and suffering before a decent standard of living was achieved, the colonists, including my grandparents, must have spent some sleepless nights pondering the dilemma of facing life under the Czar in the future or migrating to the United States. Most of them opted to remain in Russia. My ancestors elected to leave family and friends behind to seek a new life in the United States for the following reasons:

1. Loss of self government.
2. Hostility of Russian people.
3. Shortage of land.
4. Compulsory military training.
5. Preservation of their catholic religion.

While military service was dreaded, it was not a violation of conscience as was the case with the Mennonite Germans. Many of the Catholics remaining in Russia looked upon the military conscription and the hardships met by those arriving in the New World and chose the military service as the lessor of two evils. My grandfathers' catholic religion was such a part of their lives that anything jeopardizing their faith was something they could not accept. In the case of my Miller grandpa, their departure for the U.S. was delayed until a member of the group had completed his six years in the military.

In the spring of 1874, about three thousand colonists, representing the Catholic Colonies, met at Herzog to discuss possible settlement sites in Brazil and in Nebraska of the United States. The Homestead Act signed by President Lincoln in 1862, which offered 160 acres of land to any migrant that would settle on the land and the availability of millions of acres of land the United States donated to railroad companies to induce them to build the railroads, appealed to the successful farmers of the Volga region. Five delegates were selected to visit America in search of suitable settlement places. They set sail from Hamburg in April 1874 for New York. Ten days were spent in the United States but only one day at Sutton, Nebraska. With samples of U.S. Currency, samples of soil and prairie grass they returned to their homes on the Volga. In December 1874 two other delegates were sent on a similar mission. They came to Topeka, Kansas and traveled on the Atchison, Topeka and Santa Fe Railroad to Larned, Kansas. They spent about a week in Kansas and returning home, reported unfavorably of the new country, thus deterring a number of colonists from emigrating.

Despite the unfavorable expedition reports, the first group to migrate to the United States left Russia October 22, 1875, another group followed October 24, 1875 and a large group from the western side of the Volga, including my Schmidt grandpa left Saratov June 23, 1876. The largest single group to leave Russia for America comprised 108 families and started from Saratov on July 8, 1876. This party obtained passports only after paying eighteen rubles per person.

Caravans of wagons pulled by horses brought my grandpa and his group to Saratov. Pfiefer was located about a hundred miles southwest of Saratov and that meant almost a week of travel by wagon to get to the rail junction at Saratov. They were tired as the train pulled out of Saratov but the praying of the rosary was a daily routine on their entire trip. Since there were no rest rooms or running water in the coaches, each family carried a chamber pot and a can of water. The water can had a spout on it like a coffee pot. The water can was filled at the railroad stations and the chamber pots were emptied out the train windows. Lamps used in the train were oil lanterns. Coaches were often without light.

My grandfather's party traveled by train from Saratov to Hamburg, Germany and it required the better part of a week. During the trip they seldom spent a night in a hotel. The railroad started at Saratov, stopped at Tambrow, Koslow, Grjasi and Orel. At the latter station tickets were collected and checked and the passengers transferred to another train. From Orel the journey continued to Smolensk and Witebsk, a Polish city and Duenaburg the last stop in Russian territory. A large iron bridge, guarded by Russian soldiers spanned the small river separating Russia from Germany. In Germany the train stopped at Eydtkuhnen, Bismark and finally at Hamburg, Germany, where a ship of the Hamburg American Line was waiting to take them to America. At Grjasi, Russia, the party saw for the first time large piles of black and shiny anthracite coal used in the locomotives. At Witebsk, Poland, those with adult tickets had to pay 150 rubles, those with half tickets paid 75 rubles and those with a quarter ticket paid 38 rubles.

The journey over the Atlantic, stopping in England, took two or three weeks. When the sea was calm it was a rather enjoyable undertaking. On the other hand, during storms many became ill and lack of ventilation in the steerage areas made their journey most uncomfortable. From New York the party traveled by train to Topeka, Kansas, and to Hays, Kansas arriving there July 26, 1876. Grandpa Schmidt established residence in Pfeifer, Kansas, a small town about fifteen miles southeast of Hays, Kansas. Grandpa was born in Pfeifer, Russia, October 8, 1832 and died May 3, 1913 in Liebenthal, Kansas. My father George Schmidt was born in Pfeifer, Kansas, August 14, 1877 about a year after the family arrived from Russia, and died September 6, 1965.

As mentioned previously, my maternal grandpa, John Miller, also migrated from the Volga region but he and his family did not arrive in Hays, Kansas, until December 14, 1891. By then Russia was not granting visas for emigration to America. The family delayed emigration until a member of the group had completed his six years of military training. John Miller was probably John Mueller before he reached the U.S. immigration authorities. The German pronunciation of Mueller would sound like Miller so the authorities recorded Miller as his legal name. My mother Paulina Miller was born in Russia September 20, 1882 and entered the United States at the age of eight. The Miller children at the time numbered four girls and mother was second oldest. About the only recollection mother had about their exodus from Russia was that the family left Russia under darkness, that they walked a long ways and then waded across some stream. She also thought grandpa and the group that came with him had to pay the Russian authorities. As the early groups became settled, they continued to correspond with relatives left behind. In some cases they were able to accumulate funds in the U.S. to assist others migrate to this country.

My grandparents made the necessary sacrifices and migrated to the United States. Since most of the Volga Germans remained in Russia, one cannot help but wonder what happened to those people? Very little information has been released by the Russian Government regarding these people but through correspondence between relatives we know that many should have joined our grandparents and left Russia.

According to those letters, the Volga Germans in Russia after 1900 suffered indescribable hardships. As frictions grew between Russia and Germany, the Volga Germans, isolated from what they regarded as their homeland, were regarded by the Russians as "the enemy within". When World War I broke out in the Balkan States July 28, 1914, Russia and Germany found themselves on opposite sides of a war that engulfed most of the world. The use of the German language in the villages was outlawed and the villages were renamed with Russian names. Seeing these changes as a premonition of what was to come, the estimated 600,000 Volga Germans on April 25, 1917 formed the Organization of Volga Germans. Unfortunately, it also marked the beginning of the Bolshevik Revolution on November 7, 1917. The people suffered unspeakable persecutions at the hands of the Bolsheviks. Bands of Bolsheviks ravaged the land and killed priests. Correspondence was hampered and prayer was outlawed. As if these people had not suffered enough, great famines in the early 1920's and again in the early 1930's decimated the population. It has been estimated that 166,000 Volga Germans, including 60,000 children starved to death in the 1920-1922 famine alone.

On February 20, 1924, Lenin established the socialist Soviet Volga German Russian Republic. It restored some hope but it was quickly shattered. Stalin succeeded Lenin in 1926 whereupon the people were deprived of all rights, property was expropriated and agriculture was collectivized. People were arrested and thousands of intellectuals, priests, teachers, doctors, officials, both German and Russian were exterminated. In 1939 Russia and Germany found themselves on opposite sides in World War II. On August 28, 1941, Stalin abolished the Volga German Russian Republic, labeled all Germans traitors and ordered the deportation of 379,000 Volga Germans to Asia. The evacuation began September 13, 1941 as people were packed into box cars like cattle and shipped 2,790 miles to the east. Many died on the trip and legions of Germans suffered disease, starvation and death in Stalin's labor camps. □

CHAPTER III

Life In The New World

My people left a civilized community in Russia because it was in jeopardy and elected instead to tame the wilderness in Ellis and Rush County, Kansas. They faced the same problems and hardships their ancestors faced when they arrived in Russia one hundred and nine years before. This chapter explains how they proceeded to establish their colonies as they had done in Russia.

WHEN Dr. Brewster H. Higley wrote the words to "Home on the Range" on the bank of Beaver Creek in Smith County in 1873, he was writing about a fading Kansas scene. In the western one-third of Kansas, a few buffalo still roamed, but around Hays the buffalo had virtually disappeared. Kansas had become a state January 29, 1861. Ellis County had been created by an act of the Kansas Legislature on February 26, 1867 and Hays was officially incorporated November 23, 1867. By January 1876, Kansas had 130 newspapers, 379 churches, 43 breweries and 26 national banks. Thomas A. Osborn was governor.

The German settlers came to America trying to duplicate what they left in Russia, only here they would be free. It is interesting to note, even the sites they selected on both sides of the Smoky River resembled their settlements on both sides of the Volga. And as their forefathers found the steppes in Russia, the land was unimproved and covered with prairie grass. This time, however, they could relate to their ancestors' solutions to making a living with such meager resources. As their ancestors had learned, one of the first problems was shelter. Some resorted to the dugout but others built sod houses. The house was built out of prairie sod, cut into blocks and placed one on top of the other to form walls about twenty inches thick. The roof consisted of a central support called a ridge pole, tree branches were used for rafters and these were covered with plain boards, dirt and sod. The unreliability of the sod roof probably inspired this early country song:

> My house is constructed of natural soil;
> The walls are erected according to Hoyle:
> The roof has no pitch but is level and plain;
> And I never get wet till it happens to rain.

Religious devotions remained a permanent part of every family. To this day a beautiful stone catholic church stands near the center of the village. One of the most elaborate is the one in Herzog (Victoria) east of Hays. It is said the marble columns of that church were imported and when they arrived on flat railroad cars the parishioners discovered the standard wagon was not designed to haul such a heavy load. They finally used the running gear off an old threshing machine, hitched horses to it and in that manner moved them to Victoria. Blocks and tackles with ropes provided the lifting power to install the columns in the church. The church bells still ring at noon and in the evening at six o'clock. In early days every person dropped what he or she was doing and recited the "Angelus". Attending mass on Sunday and Holy days was a serious obligation for every member of the church. Some of the problems encountered by the Christians along the Volga repeated when the early settlers came to America. They arranged for churches but often experienced a shortage of priests. At times a priest would visit a village once a month.

While the Volga Germans labored long and hard to provide a livelihood for their families and left no stone unturned to instill in their children the faith that supported them for so long, they failed miserably in not recognizing the need for public education. Most of the immigrants could not read or write. Girls were virtually ignored so far as teaching them to read and write. Furthermore, any education they received was in the German language. Not surprisingly, this attitude also followed them to the United States. English was not taught in their parochial schools until the state demanded it by the Kansas School Attendance Law of 1874 which required twelve weeks of yearly attendance at a public or private school for all children between the ages of eight and fourteen. Public school districts were organized to comply with the law. Father

Stollenwork, pastor at Liebenthal from 1899 to 1931, had these comments regarding education:

> *Special stress was laid on religious instructions. The catechism was taught and explained always in German. Progress was slow owing to several causes. In the first place attendance was irregular and intermittent. Many did not then realize the necessity of education, being accustomed to the village life in the old country and not coming in contact much with the outside world. Their business dealings were almost exclusively with people of their own language and faith. Another reason was the work on the farm which caused the parents to keep (especially the older children) out of school as long as possible and to withdraw them as soon as spring work commenced. The language of the parents was German, and the children therefore conversed in that tongue both at home, on the street and on the school ground. 'Let the children learn the German first; they can learn the English when they grow up' was the sentiment expressed when the older men discussed school matters.*

This probably explains why my father could neither read nor write. He could sign his name in English. Since he was the oldest boy in the family, he probably received little education and even that was in German. Parochial schools delayed complying with the new law as long as possible. German was taught in the morning and English in the afternoon. Mother received little or no schooling but by individual effort she learned to read and write in German. She would bring arm loads of German books from the Catholic College Library in Hays.

Living in the villages and farming outside the village was the custom of the Volga Germans while in Russia. When the family grew or new settlers arrived a new village was started. So it was natural for them to repeat the process here in the United States. The original settlers established a number of villages with the catholic church near the center of the town and homes surrounding the church. Farming was done in the country outside the village. My Miller grandfather, who came to America about fifteen years after my Schmidt grandfather, and other immigrants from Russia proceeded to establish another village about ten miles northwest of Hays, Kansas. He and other German immigrants, plus a number of Irish catholics had acquired land in that area. Dad purchased 160 acres in 1904 at a cost of eleven dollars per acre and another 160 acres a little later at a cost of twelve dollars per acre. Dad's land was purchased from a man by the name of Strailey who had homesteaded under Chester A. Arthur, President. There was a building on one of the quarters but really not suitable as a house. A school district had been organized and a school built four and one-half miles from the old house on the farm. Grandpa Miller's land was adjacent to our farm.

The public school was located on high ground between the south slope of the Saline River and the north slope of Big Creek. My grandpa Miller and a devout catholic Irishman by the name of James Ross were actively involved in a movement to build a church near the school. It was to become the site of another village. John O'laughlin donated a tract of twenty acres for the church. A frame church was completed on this ground in three months by about twelve families working together and the church was solemnly blessed by Bishop Cunningham of Concordia November 29, 1906. Dad bought a small tract of land near the church and built a two room frame house on it. He also built a makeshift barn out of tree timbers, straw, etc. to house horses in the winter for the trip to the farm. The name of Hyacinth was probably attached to the area in honor of Reverend Hyacinth, first provincial of the Capuchin Order of the Pennsylvania Province. Grandpa Miller sold his farm to James Ross and constructed a building north of our house that became his home and the first store in the village of Hyacinth. During 1913 and 1914 a parochial school was built providing space for two school rooms and six rooms as living quarters for catholic nuns who would teach the children. The Diocese was never able to supply the nuns and the building never served the purpose for which it was constructed. I and all of my brothers and sisters attended the public grade school in Hyacinth.

The children of the original settlers adopted the American way of farming and built their homes on the farm and not in the village. My father saw the impracticality of the village arrangement, moved the two-room house to the farm and added to the size of the building. Grandpa's store was an economic failure and he moved to Hays. The parochial school was torn down and eventually the same fate befell the church. There were other second-generation villages in Ellis and Rush counties that fared little better. □

CHAPTER IV

MEMBER OF A LARGE FARM FAMILY

It seemed like the large family and the farm were not only compatible but essential to the farm operation of the times. We are speaking of the period prior to World War I and the end of World War II. This chapter presents an authentic description of life on the farm during this period, as well as various farming operations in vogue at the time. Many of the labor saving devices on modern farms today were not available then. The chapter reveals how cruel mother nature can be by destroying the fruits of one-years labor in minutes.

LIEBENTHAL, Kansas was a rather small village and the probability of Paulina Miller meeting George Schmidt was excellent. Grandpa Schmidt had four sons about the same age as Grandpa Miller's six daughters. They were married November 27, 1899. I have found no written record of the wedding feast but it probably followed the custom of the day.

Marriages were not made in heaven as proclaimed. In most instances the marriage was arranged by the parents of the couple planning to marry. Strict discipline in the family, religious education stressing respect for parents and a committment to obedience resulted in the young not only seeking the approval of their parents but their blessing. No other event was so important to the families involved as was the marriage feast. It became a community event. Once the marriage was approved and the wedding date selected, the couple had to contact their parish priest. A wedding in the church could not take place until the banns of matrimony were announced three Sundays before the wedding date.

With those preliminaries out of the way, the time arrived for the parents to plan and prepare for the wedding feast. Those wedding feasts would last for two or three days. Depending upon facilities available, the wedding feast might be held in either the bridegroom's home or in the bride's home. Invitations had to be extended either in person or in writing. Food and refreshments had to be prepared. A band was engaged to provide music for dancing. Material had to be purchased for the bride's dress and the sewing job delegated to someone. Cooking had to be delegated to someone and the baking was done in a mud stove.

After receiving their parents blessing, they probably walked to the church on their wedding day and the marriage vows were exchanged during high mass. It was understood that the bridal party would be detained as they left the church by the altar boys pulling a rope across the exit. After one of the best men tipped the boys the party was permitted to leave. The eating, drinking and dancing followed the church wedding. In Mom and Dad's wedding the music was probably furnished by local talent using first and second violins, dulcimer and cello. As the bride and groom started dancing, guests pinned paper currency on the bride's dress and veil. I can't imagine a German wedding without group singing and some card playing.

Mom and Dad set up housekeeping and farming on a small tract of land south of Liebenthal. Dad's farm equipment consisted of three horses, a wagon and a plow. Since the land was not far from Grandpa's place, he probably borrowed some equipment from home. My two oldest brothers were born during their stay at this place. Bernard (Ben) was born May 16, 1902 and Henry was born September 19, 1903.

With the purchase of land northwest of Hays, the family moved to the farm in Ellis County and later built the two-room house in Hyacinth as a home for mother and the children during the school year. Ben and Henry enjoyed the home in Hyacinth because it was only about a hundred yards from Grandpa's store. Dad would commute to the farm in the winter to take care of livestock. The family lived on the farm in the summer. Three of us made our appearance while the folks had the house in Hyacinth: Regina April 18, 1905, Marie November 1, 1906 and Walter August 24, 1910.

The severe winter of 1912 convinced the folks to move the two-room building to the farm and to build an addition to it for the larger family. The old house acquired with the land was moved and became a granary. The building was moved from Hyacinth over four and one-half miles of rolling terrain by horsepower. Additions to the house included a large dining room, another bedroom, a kitchen, a screened back porch and a clothes closet. A stairway led to the floored upstairs that remained unfinished. This large area became the boys dormitory. It was cold in the winter and hot in the summer. Mother solved our winter problem by equipping each bed with a feather quilt about eight inches thick. It was not uncommon to wake up and find a layer of snow on top of the quilt and find some in the lamp chimney. Sleeping outdoors provided some relief in the summer.

Dr. Middlekauf showed up from time to time to bring the remainder of my brothers and sisters. Cyril March 21, 1912, Elizabeth March 6, 1914, Raymond (Ray) August 7, 1915, Eleanora July 29, 1917 and Gilbert (Gil) September 12, 1920. In the nineteen eighties this would be a large family indeed. Ten children in all, four older than I and five younger. In the early nineteen hundreds there were a number of farm families in our neighborhood with more than ten children. Children were an asset on the farm and an integral part of the business. Much of our food was produced on the farm: Meat included beef, pork, turkey, goose, duck and chicken. Vegetables included potatoes, radishes, onions, beets, cabbage, lettuce, asparagus, beans, etc. A good supply of milk provided butter and cheese; Eggs were plentiful.

As time and money permitted, a well was drilled near the house, a storm cellar and cream house combination was constructed of stone, a stone chicken house was built and a dug well was equipped with a windmill. Unfortunately, the well near the house supplied only enough water for household use so water for the laundry came from the dug well.

In 1920 Dad bought a 1920 Model T Ford car. It was four-door, black in color and had a black canvass top. To house the new car a frame garage was built. My older brothers mastered the skills required to drive the car but Dad encountered some difficulty. One day he came to pick me up at uncle Fred Ross's house about two miles from our home. He parked the car in front of the house hoping to exit by making a circle to the left. He evidently misjudged the distance necessary and failed to apply the brake soon enough as the car came in contact with a barbed wire fence inflicting some nasty scratches on his new car.

Believe it or not, I have three Irish uncles and they are from the same family. George, Charles and Fred Ross married Mary, Charlotte and Elizabeth Miller. Probably not a record but certainly unusual.

What was it like to grow up on a farm without electricity, TV, telephone, radio, no skate board, no purchased ball or bat? The most enjoyable play for me as a youngster was activity that I and my brothers concocted through our imagination and our pretending to be something or someone else. I can recall sitting behind mother's wood cook-range, where it was warm, playing with my toy horse. (Mom's sewing scissors). When the weather warmed in the spring, the snow melted and formed little streams to the roadway. We constructed barns on the sunnyside of the road bank by digging into the bank with a stick or metal object (perhaps some of Dad's tools or mother's utensils if we didn't get caught) and then we would sculpture the horses we needed to do the farming from mud along the little streams. There was a distinct advantage to playing at that location. The distance from the house was such that mother was never sure whether we heard her call for supper or whether we were simply ignoring her call.

We did not live near a creek or river but excess water from the stock tanks flowed into a small pond that provided a convenient wading pool, albeit muddy. Along that pond three cottonwood trees flourished and became quite tall. What can be more fun for a kid than to climb into the top of these trees in the summer and enjoy the summer breezes as the branch on which you are sitting sways back and forth? There is no noise to interrupt your dreams up there.

There was no sand pile to play in but I had something much better. The granary usually had a large mound of wheat in it, the granary roof provided shade and with the granary door open, the Kansas wind provided sufficient ventilation. What kid would not take his shoes off and play in a pile of wheat, much cleaner than a pile of sand?

While we did not have league baseball equipment, we played baseball in our neighbor's meadow not far from the house. Any piece of soft wood could be sculptured into a bat and all that was needed for a ball was string conserved from purchases at the store.

While we had no manufactured toys, we had something better, live toys. In the spring of the year we played with baby jack rabbits, baby chicks, baby ducks, baby pigs, baby puppies, baby kittens, baby calves and baby colts. The baby jack rabbits had to be captured in the field and if we were lucky we caught one of those little white pigs as it strayed from its mother. We spent very little time in the house in the summer. There was no fear of a child getting kidnapped and as youngsters we remained within calling distance. However, mother told us how startled she was when it appeared that her young one had disappeared. Mom and Dad went to the field to look at a fine crop of wheat. Ben was not very tall at the time and his constant companion was a St. Bernard dog. They followed a path into the wheat field with Ben and the dog following. After a time, mother turned to see if Ben was alright but Ben and the dog had disappeared. He and the dog had strayed into the wheat, which was taller than either of them, and kept walking away from the folks. They kept calling but there was no response. Things were getting serious as time passed and the boy was nowhere to be seen. Finally, they caught sight of the dog appearing above the wheat as he stood on his hind legs. Walking over to where the dog appeared they found Ben and the dog safe and sound.

The farm family was a close knit organization. Certainly there were differences of opinion and at times dissention between members but the family remained one unit. We worked together, we sang together, we ate together and we prayed together. The first sound we heard in the morning was dad leading the family in morning prayer. The family was not a democracy. It was understood that dad was the boss. The only judge we could approach for redress of grievances was mother and hope that she would be successful in pleading our case before Dad. As soon as we were mature enough to take care of a job, we were assigned one.

Everyone had to learn how to milk a cow. Cows were milked twice daily. Sometimes we had a dozen or more milk cows. The milking machine was not in existence and even if it had been there was no electricity to power the machine. With so many animals around, providing food and water for them was a job to be assigned to someone. After the milking was done, the cream had to be separated from the milk using a hand operated cream separator. The cream was accumulated in a cream can and the skimmed milk was fed to the dogs, cats, pigs and chickens. Cream not needed for family use was taken to Hays and sold. After washing the kitchen dishes, someone inherited the job of washing the tank and the separator parts through which the milk passed. The chicken house contained nests built about three feet off the ground so hens would have a place to lay eggs. Hens are an independent lot and frequently find a nest they prefer and will deposit their eggs elsewhere. So searching for and gathering eggs each evening became a chore for someone. Eggs not needed for family consumption were accumulated in egg cases and taken to Hays to be sold. Another chore that I remember took place in early spring. Land had been surveyed and easements were granted for roadways in the rural area but very little had been done toward road construction. Many easements remained pastures with perhaps a wagon trail through it. Fences had been constructed to keep livestock from straying, so the easement was not pastured, providing desirable grazing in the spring. One of my jobs was to open the gates and let the cattle graze on the easements surrounding the farm. The cattle had to be kept together, more or less, and I had to guard that none strayed into the neighbor's field. Those were carefree days for the most part. It was relaxing to lie in the grass and watch the fleecy white clouds sail overhead. It was also a good place to read or study. In fact that is where I memorized my latin responses as an altar boy. The priest would open the mass at the foot of the alter by saying, "Intribo ad altare dei" and I would answer, "Ad deum qui laetificat juventutem meam." Translation, "I will go unto the altar of God." "To God who giveth joy to my youth."

There was always the job of taking care of the setting hens and the little chicks after being hatched. Sometimes baby chicks were purchased and put under a kerosene brooder. We raised hundreds of baby chicks. The male chicks became fryers for the table while the female chicks became laying hens as the older hens were culled and sold in Hays.

While some of our neighbors assigned field work to both boys and girls of the family, that was not the policy in our family. With a family of ten and no more conveniences than mother had, it became crucial that the girls help her with the cooking, dishwashing, laundry, ironing, sewing, mending, canning, house cleaning, etc. There were times when all of us boys were in the field, during harvest for instance, when my sisters did the milking and took care of the milk. Generally speaking, the outside work was boys' work and the inside work was girls' work.

Most of my life on the farm was before we had tractors, trucks, elevators, combines, etc. Since our farm had about 200 acres of cultivated land it required many work horses and much horse drawn machinery. For tilling the soil we had the plow, harrow, cultivator, drill and the disk. For harvesting wheat we had the header and the header box. For feed crops we had the mower and the rake. While we had a single row binder for the row crops we did not use the wheat binder. Most of the work horses were raised on the farm and broke to work when they reached the age of three or four years. Some of the horses received preferential treatment because we selected them as our saddle horses. Mine was a little black mare with a broad white stripe on its forehead and one white front foot. Riding horses was a very popular mode of transportation on every farm.

Starting with ground preparation for the planting of wheat and ending with the harvesting of the spring crops, let us take a look at what was involved in the way of field work. If rains came in July and August, sufficient to soak the soil to a depth of five or more inches, the tool used was the single or double mold-board plow. We used five horses to pull a fourteen inch gang plow, four horses to pull a twelve inch gang plow and three horses to pull a sixteen inch single plow. Sometimes we had four of these plows in the field at the same time. This means someone has to rise at daybreak, bring fifteen or twenty horses to the barn, feed the horses and harness them before breakfast. We had a grain bin in the hay-loft of the barn with a spout that allowed us to lower grain into a bucket by opening a sliding door. Each horse received his portion of the grain and had access to roughage such as oats, barley or wheat. We had one horse intelligent enough to help himself to some grain during the night by pulling the sliding door on the spout with his teeth and allowing the grain to drop to the floor of the barn.

After breakfast four of us had to get our three, four and five horse teams organized by putting bridles on them and hitching them together. The teams were then walked to the field and hitched to the plow. We hoped to get started by seven o'clock in the morning. A two-hour break was taken at noon when the horses were brought back to the barn for water and food. Starting time after lunch was two o'clock and seven o'clock in the evening was quitting time. The horses were brought back to the barn, unharnessed and fed before dinner. I don't recall how old I was the first time I held the reins for three horses pulling a plow but I can assure you it was before I was ten years old. Keep in mind there were two older brothers and Dad in the crew keeping an eye on my progress. When the rains stopped and the ground was too dry to plow, six horses were hitched to a disk and it became the tilling tool.

Wheat planting took place in September or October and the red hard winter wheat brought from Russia was used. A wagon was backed up to the granary door and loaded with seed wheat, which had passed through a sieve, making certain no foreign material was in the wheat to interrupt the flow of wheat through the drills. The drill was a machine with two wheels, a grain box, a set of disks that opened the ground, a conveyor system that dropped the seed into the ground and a set of chains that covered the seed after it was planted. We had two or three drills in various sizes. Each required four or six horses to pull the machine depending upon the size. Keeping seed in the drill box was accomplished by using buckets or old wash kettles to carry the seed from the wagon to the drills. Since all of us attended public school, which opened in September, there were times when the older boys came to school after wheat planting was completed.

Not all of the cultivated ground was ever planted entirely to wheat. Feed crops were planted in the spring and harvested in the fall for livestock feed in the winter when pastures were not available. Feed seeds were either planted by a lister, which produced a single row of feed, or broadcast which produced a mass of feed. The first was harvested in the fall by binding and the latter was mowed and raked in the fall. The bound feed was accumulated in shocks in the field by setting the bundles upright to protect the seed on the bundles. The loose feed was gathered in small stacks by forks and raking. The feed was usually hauled and stacked near the barns in the fall of the year.

Preparing the ground for spring crops did not require either the manpower or the horsepower required for the wheat crop. In years when the wheat crop did not make it through the winter, the acreage for spring crops was increased measurably by planting barley or oats. In addition to the planting of feed crops, spring was the time for the family garden. Some acreage was left near the house for that purpose. Peas and potatoes were some of the first seeds put in the ground. A plow was used for opening trenches for the potatoes in the spring and a plow was used to dig the potatoes in the fall. The other garden crops were planted by hand.

Probably the most labor intensive work on the farm occurred when the wheat was harvested. It was the only time of the year that Dad employed outside help. There were a number of times when neighbors helped each other in the event of emergencies such as tornadoes, etc. I remember mother insisted that all of us seek shelter, when an ominous storm came up one night and for some reason we could not use the storm cellar but the new ice cave was finished so we used it for shelter. Next morning we learned that a tornado missed us by about four miles killing a number of people and devastating the buildings and property of one of my uncles, Walt Stanton. He and his family were saved in a storm cellar but his buildings and personal property were demolished.

The harvest equipment used was a header and a header box. The header was a cutting machine mounted on a tricycle that cut the heads of wheat and conveyed them up an incline into a big box mounted on a wagon. The header was pushed by six horses and the driver stood behind the horses on a platform and guided the header by turning the third wheel of the tricycle. The header box was pulled by two horses and had to accompany the header as it moved around the field. If it fell behind the speed of the header or moved faster than the header the heads of wheat fell to the ground. The object was to accumulate the heads of wheat in the box to deliver them to a suitable site where the wheat heads would be stacked. A standard harvest crew consisted of the header driver, the kid that drove the teams pulling the header box, four men to load and unload two header boxes and Dad who stacked the wheat. As a youngster I drove the teams for both header boxes. When one box was loaded, they hoisted me to the top of the header elevator and when the next header box arrived under the elevator I would jump off. As I matured my assignment was header driver.

Wheat heads were normally accumulated in pairs of stacks. A stack was about eight or ten feet wide and twenty or thirty feet long. As the stack rose above the height of the header box it would start tapering toward the center. Dad's instructions to us, when we were stacking was, "Keep the center full." In other words the idea was to stack the wheat so the stack would shed water in the event of rain. The height was limited by the reach of the unloading crew using manual forks. The purpose of the two stacks, parallel and far enough apart to allow the header box to enter, was to allow room for the feeder and the threshing machine to enter. Sometimes Dad was able to hire good harvest hands and sometimes not so good. I recall an Irish hand told us he had traveled through a little German town in Illinois when all of a sudden the front door of the local saloon flew open and men shouting, "Raus mit em" were throwing a guy out of the saloon. We told him it meant, "Out with him." Harvest brought a heavy burden on mother and my sisters. Not only did they have to cook and house more people, they frequently took care of other chores because the boys were in the field harvesting.

The most exciting time of the year was the threshing season. This part of the harvest was performed by someone in the business of threshing on a contract basis. The contract fee was based on a price per bushel and I don't recall Dad ever signing a written contract. In those days, business in general was conducted by oral committments, a man's word was his bond. Before the thresher arrived, there was another little chore Dad thought needed doing. Winds in western Kansas can blow a lot of wheat heads off the stack before the heads have had time to settle. One of my jobs was to remove the heads that had blown between the stack and place them at the end of the stack so when the threshing crew pulls its equipment between the stacks the wheels will not squash a lot of wheat into the ground. While doing that, one of the many hazards lurking around a farm made itself known to me. As I was about to pick up a forkful of wheat, I heard a strange rattle coming from the stack. Backing up a step or two, I swung my fork into the side of the stack and out pounced a rattle snake. Once the rattler was on the ground, it lost all the advantage in the battle for survival.

The thresher would arrive with the steam or deisel engine pulling the separator and the feeder. Behind that would come the cook-shack, the water-wagon and the coal-wagon. The thresher provided the men, housing and cooking for the men, provided the fuel and maintained his equipment. A normal crew consisted of an engineer, a separator man, a straw stacker, about eight fork men, a water man and two or three cooks. It was our job to take care of the grain dumped into the wagon parked next to the separator. The thresher would drive his engine pulling the separator between the stacks until the separator was in the correct position. The separator was then leveled by lowering some of the wheels of the separator into little trenches dug with a spade. Blocks were placed in front of the separator wheels and the engine was detached from the separator. When the engine had cleared the stacks, members of the crew pushed a feeder between the stacks and attached it securely to the separator. The engine pulled fifteen or twenty yards from the stacks, turned around to face the separator and aligned the pulley of the engine with the belt pulley on the separator. The next job was putting the belt on the pulleys and tightening the belt. With a toot of the steam whistle and a puff of smoke, the equipment started humming and the crew started pitching wheat onto the feeder. The separator dropped the wheat, a bushel at a time, into the wagon and carried the straw on a lath conveyor to the straw stack.

Our job was to haul the wheat from the field and scoop it into the granary. Pitching wheat heads into the feeder or scooping the grain into the granary through a window above one's head was a backbreaking job.

This was the scene on the farm when we had a crop. There were many years on the farm when all the money, time and hard work resulted in no crop. I can remember one summer when a bad storm arrived from the northwest and when we came out of the storm cellar, hail the size of golf balls had accumulated to a depth of several inches. We had a good stand of wheat about ready to harvest. When we walked out to the field after the storm and found the wheat lying flat on the ground both Mom and Dad wept. There were other years when rains did not come in time and March winds literally blew the young wheat plants out of the ground. Books have been written about the great dust storms of the 1930's. Wind erosion blew good top soil into drifts along fence lines and around buildings to mounds as high as ten feet. Dust clouds would turn day into night.

The hard work and the long hours of work on the farm were concentrated in the summer months. Once the wheat was planted and the feed put up, getting the kids to school and doing the chores became the daily routine. We had one tough job for a time in the winter, namely, putting up ice. In addition to the storm cellar, we dug a rather deep hole in the ground, drilled holes in the corners at the bottom of the hole and built a roof over it. This was our ice cave. At the time, Big Creek about three miles south, carried a stream of water sufficient to provide us with blocks of ice from six to eighteen inches thick. Putting up ice always came during the coldest part of the winter. Usually around Christmas and New Years. At daybreak three or four wagons and our work crew would drive to the creek, mark and cut the ice into blocks about 18″ x 24″ in size. The thickness depended on mother nature. When we had enough blocks to fill the wagons, the blocks were loaded for delivery to the ice cave. One day my brother Cyril meandered too close to the cutting edge of the ice and fell into the ice water. He was pulled out quickly by members of the crew and it didn't take long before Cyril had a complete armor of ice. We usually got home after dark, unhitched the horses and ate dinner before unloading the wagons. After dinner the wagons were rolled up to the door of the ice cave and the blocks were stacked in the cave by sliding the blocks down a wooden shoot using ropes and metal hooks. Space was allowed between the ice and the walls of the cave so straw could be packed into the space when the cave was full. By so doing we had ice all summer for the ice-box in the house and for other purposes. The drying of Big Creek and the warmer winters ended this source of ice years later.

In my work over the last few decades, I have been asked by people much younger than I, "What do you think people will do if we have another depression like we had in the thirties?" My standard answer was, "You will do just what we did, whatever is necessary to survive." While the stock market crash occurred in 1929, the real effect of the depression hit the farm about three years later. We had increased the farm by leasing some land and used modern machinery to produce good yields but the deflation of agricultural prices put many a farmer out of business. I personally delivered a load of wheat to market for Dad and he received twenty-two cents per bushel for the wheat. I accompanied my brother-in-law as he delivered a wagon load of ear-corn to market and received nine cents per bushel. That load of corn, hauled about ten miles to market, brought that man $2.34, enough to pay for a sack of bran for the milk cow, a sack of flour and some onions for the family table. Yes, I saw poverty in the thirties. Dad saved his land but only because he was successful in obtaining refinancing through the newly created Federal Land Bank of Wichita, Kansas. While we had little or no money to spend, we never went to bed hungry nor were we ever cold. When Dad passed away September 6, 1965, he left his children an estate worth $50,000.00. Not bad for a man who could neither read nor write the English language.

An issue of the Kansas City Times of March 18, 1986 carries this headline, "CONGRESSIONAL STUDY PREDICTS END OF 1 MILLION FARMS". Hundreds of thousands of farm families were forced to leave their way of life during the Great Depression. Is it possible that history is repeating itself fifty years later? The media has been reporting a number of suicides among farmers in financial straits. I read an article lately written by a psychologist familiar with the rural problems and I agree with his interpretation of the dilemma. The loss of the farm is not the loss of a job to the farm family, it is a loss of part of their lives. No one will ever make me believe that my parents chose farming to make a lot of money. They chose farming because they loved the land and it was regarded as the best place to raise their children, to teach them the value of honest labor, teach them the ten commandments and other values our parents cherished without the distractions and peer pressure of the cities. If the family is the foundation of our democracy, then I would like to suggest the farm as the best place to keep the family together. Is farming just another industry or is it a desirable place for millions of people to live, raise their families while at the same time producing enough food to feed most of the world?

When the six of us brothers were at home, there were many times when one or more of us worked for neighbors. I recall once when we had no harvest but Steve Sack near Big Creek had a good crop and asked us to do some custom harvesting for him. Machinery, horses and crew moved to the field. I had my usual job, driving the header box teams. I recall how embarrassed I was one noon when the crew left for the field without me. Steve had a small cherry orchard and the trees were loaded with cherries. While the crew had lunch and the horses were feeding, I slipped out to the orchard to pick some cherries and only when the cooks came out to pick some cherries, did I realize the crew had gone to the field without me.

Unlike a corporation, the family organization does not live in perpetuity. By the time I graduated from grade school, my older brothers and sisters were married and had left home. Ben delivered petroleum products to farmers and the other three were involved in farming on their own. As the depression wore on, one by one gathered his or her belongings and moved to Chicago seeking employment. Only Marie and her husband returned to Kansas. Unlike their ancestors, the children did not marry and move in with their parents. My oldest brother and his wife lived with us for a short time, making thirteen around the table. Some land was leased about forty miles west of Hays to put out a crop for the new couple. The venture must not have been very successful because it was never repeated for any of the other children. The farm was near Collyer, Kansas and I remember it as the place where the cat ate the lamp. When the crew arrived at location, we discovered we failed to bring the kerosene lamp. Dad, using his ingenuity, poured some lard into a tin and made a wick out of some material. The lamp worked but the shack had no doors and during the night a cat devoured our lamp.

With the two older brothers gone, I was the oldest boy on the family farm and much was expected of me at the ripe age of fourteen. It was also the period when the tractor replaced the horses and the combine replaced the header. Farming had never been very profitable but no one foresaw the bleak future ahead of us. In addition to the mortgage debt on the farm, there were other obligations around town. Since I was the oldest, I was invited to drive Dad to town to negotiate for loans and repayments. Dad would say, "When I didn't need a loan the banker would stop me on the sidewalk and offer to make me a loan. Now when I really need a loan the same banker won't approve a loan." In one place he received this bit of wisdom, "A man should work for money until he reaches fifty and then his money should work for him." A lot of help that was to Dad. At another place Dad wanted to borrow money and he was projecting where the resources would be created to make repayment of the loan. Joe Schaefer listened politely to his financial forecast and replied in German, "George you are shooting too far. You will never hit." I kept a rudimentary set of books for the farm and the results did not look good even to a grade school kid. Another trip with Dad was to purchase a tractor. Brother Ben supplied gasoline to a number of farmers and had knowledge about different makes of tractors. Allis-Chalmers was his recommendation. I went with Dad to the dealer in Hays and the dealer, with perspiration on his brow, was pounding Dad's ear about the quality of his product. At one point he said, "I can drink more water than this tractor uses on the hottest day of July." It must have convinced Dad as we bought the tractor at a cost of about $1,500.00. It was up to us boys to operate and maintain the tractor according to the printed instructions. We equipped the tractor with a headlight and operated the tractor night and day during plowing season.

We had some nice neighbors and some who were not so nice. One good Irishman, Pat Mahoney, lived about a half mile from us in an old shack. Pat was elderly and probably lonely so he sponsored the immigration of another bachelor, Jim Finnegan and provided Jim a home. The folks kept an eye on Pat while he was alive and on Jim thereafter. Mother told us Dad was late coming home from town so Pat offered to milk the cows and when mother went to see how Pat was getting along, the cow had kicked Pat down the hillside, milk stool and all. Jim had suffered a broken collar bone while playing football in Ireland and nothing was done to correct the injury. The folks felt sorry for Jim and when one of those Kansas blizzards came along they would invite Jim to stay with us for the night. Mother put him in the attic with us under the feather quilts. Next morning after a good breakfast, he would return to his shack. I have often said, "My folks took care of Jim's physical needs and the Ross's took care of his spiritual needs." The Ross's would see that Jim went to mass on Sunday. □

CHAPTER V

Country School and Church

Most of my cousins received their grade school and high school education in parochial schools but that was not true in our family. The parochial schools were built near the churches in the little villages. In our locality the catholic church was built near a public grade school and that is where we received our grade school education. This chapter describes how a one-teacher grade school was conducted and how limited the school and the playground facilities were.

THIS seven year old faced an entirely different environment than my children faced at seven. First of all, German was the only language spoken by my parents and English was the only language used in school. I learned English from older brothers and sisters, from my teachers and from peers in school. Despite the language problem, I am convinced bilingual education in our elementary schools is a mistake. I am glad we were forced to learn the English language in elementary school. It made life in this country easier. While I appreciate knowing and understanding some German, it has been of little value to me except on trips to Europe.

As mentioned before, our public grade school was located four and one-half miles from home. All grade student arrived at school by walking or by horse and buggy. The Al Ross children lived just north of us, on what was Grandpa Miller's home at one time and they drove a buggy to school using one white horse. White must have been in style, I remember driving two white horses pulling a spring wagon to school. Our mail box was a mile from home so that we could stop for the mail on our way home by driving through our field. Sometimes we had to lower the fence to cross and to put the fence back afterwards. We had a weekly newspaper known as CAPPER'S WEEKLY published in Topeka, Kansas. We always read the comic strip "The Intellectual Pup" in the weekly paper. The Brungardt children lived two miles from school and walked to school most of the time. The school day started at nine o'clock and ended at four o'clock. Either Dad or my older brothers had to harness the horses and have all of us aboard before eight o'clock in the morning. When it was extremely cold, the folks would heat bricks, wrap them in burlap and place them at our feet in the spring wagon. When we arrived at Hyacinth we would unhitch the horses and put them in the makeshift barn Dad had built.

The school was a wood structure with an attached entrance-way on the northside. Enrollment was about twenty or thirty students from first to eight grade. A class might have one or two students. The building was heated by a coal burning stove in the middle of the room. Upperclassmen would help the teacher bring coal from the coal-shed as needed. As I recall, we used wooden desks built for two. The desks were not adjustable so the teachers' ability to make students comfortable was limited. There was no playground equipment. We had to improvise a ball and bat to play baseball in O'Loughlin's pasture, when the weather permitted. At the end of recess, the teacher would appear in front of the school and ring a small hand bell. When the wind was carrying the sound of the bell away from us and we were in a tight baseball game, it was no small task for the teacher to get the kids back to school. Sometimes the teacher found it very difficult to stop a fist fight between two boys who towered over the teacher in height. Did the teachers slap students when they talked back to them? Indeed they did. One of the men teachers broke a small coal shovel over the rear-end of a particularly belligerent young man one morning.

Were the kids able to learn under those circumstances? They did very well, thank you. My cousins went to parochial school in Hays with hundreds of classmates and had much better facilities. I was invited to attend school with a cousin in my class and was surprised to find we were farther along in our subjects than they were. The explanation must be the fact that with smaller classes we received more personal attention from the teacher. This gives me an opportunity to mention text book material of the day. A first grader not only learned the letters of the alphabet and the words but the stories were designed to teach the child the value of work, responsibility and thrift. THE ANT AND THE CRICKET was such a story. The cricket enjoyed the summer relaxing and singing whereas the ant carried food into storage. In the winter the ant had food to eat and was warm. The cricket froze to death seeking food. There were many such stories that introduced the young to values perhaps as important as learning to know letters and words. Nevertheless, these lessons of life have been removed from children's books. I think psychologists now agree that most of our character is developed very early in life. What one is exposed to in the home or the school in those tender years will affect that person's behavior for the remainder of his or her life. I will be forever grateful for the values I was taught during those years.

We had a number of extra curricular activities. There was the box social, the Christmas play, the spelling bee and the cyphering contest. When I was in the sixth grade, I won the school cyphering contest and was second in the school spelling bee. It was quite a thrill to be standing when the eight graders had been retired.

A thought just crossed my mind as I was thinking about my elementary education. In view of the attitude of my ancestors toward public education, perhaps I owe my elementary education to the fact that Kansas mandated elementary education. There was no parochial school at Hyacinth and Hays was about nine and one-half miles away. That century of isolation in Russia seems to have convinced them that man's ultimate reward was determined by his nearness to God and his devotion to the land. While one of life's embellishments was a public education, it really wasn't that important.

My religious faith stems largely from my parents and was influenced greatly by an order of catholic priests known as Capuchen Fathers who served the Germans in Russia and also assumed spiritual charge of catholics in Kansas. With the establishment of a Catholic College in Hays, priests teaching at the college would take care of the church at Hyacinth. While we had a parish house near the church, we did not have a resident priest at St. John's. It was regarded as a mission church. Due to the proximity to Hays, we always had a priest on Sundays and Holy Days when some other parishes were not so fortunate. Five days each week we drove to Hyacinth to attend school, on Saturday we drove to Hyacinth to attend catechism and on Sunday we drove to Hyacinth to attend mass. Each of us had a catechism book and the priest would assign work for us to memorize by the following Saturday. Catechism class would last for about two hours. At times the homily during mass was delivered by the priest in English and sometimes it was delivered in German. Some of the funds for the church were raised by auctioning the pews for the following year. The pews in the back of the church always cost more rent than the pews in the front. Maintenance was furnished by the parishioners on a voluntary basis. Dad usually had the job of painting the steeple because he did not fear height. The church burned to the ground twice and each time the parishioners would pool their effort and rebuild it.

Babies were brought to the church for baptism very early in life. Uncle Walt Stanton and Aunt Mary were my godfather and godmother. Holy pictures were in about every room of our house and daily prayer was standard procedure. The family recited morning and evening prayers daily. Grace before and after meals was regular routine. Everyone was expected to attend mass on Sundays and Holy Days and to receive the sacrament of penance and communion about every two weeks. The lenten season was observed by fasting and reciting of the family rosary each evening. I was an altar boy for several years and enjoyed delaying the married couples at the front door for a tip just as my ancestors did many years before. While I pondered the idea of becoming a priest, the committment to do so was not forthcoming. While I did not select the religious vocation, some of my cousins, nephews and nieces did.

Years ago, the company I worked for sent a new accounting recruit with me by the name of Dana Flanigan. To my knowledge Dana claimed membership in no church. He made the remark once, "If there is one true church, it is either the catholic church or the jewish church, the other churches are not old enough." Depending on the definition of a church, millions of people on the islands, in the orient and in China or India might argue about the age of their church. Dana must have been thinking of people who believe in God. It has been said, "If there was no God, man would invent one." It seems like the human being existing on some remote island, forgotten in some African jungle, hidden in the mass of humanity in India or China, instinctively worships an invisible power greater than himself or herself.

There are several catholic churches in the world such as the Orthodox Catholic Church, some churches in England call themselves catholic churches and then the larger Roman Catholic Church. As a member of the Roman Catholic Church, one of my church heroes is Pope John XXIII. He started what one day may result in the scripture prediction of "One shepherd and one flock." It appears the Orthodox Church resulted from a dispute about the seat of the pope. Should it be in Rome or in Constantinople? Now the Orthodox Church adds the name of the country where it exists such as Russian, Greek, Armenian, etc. When we were reading about Martin Luther and his ninety-five theses in world history, one of my classmates remarked, "Miss McClain, this makes the protestant church a branch of the catholic church." Miss McClain, a devout protestant, replied, "If Martin Luther heard you, he would turn over in his grave." I have no idea how many hundreds of organizations claim belief in the birth, crucifixion and resurrection of Christ. In His years of teaching there was not the slightest inference that His church was a democracy subject to the whims of man. Christians remind me of some of our supreme court judges. They say, "We know what the constitution says, but we believe the founders of our country meant something else." The inscription on John Kennedy's memorial probably sums it up best. "From those, to whom much has been given, much will be expected." □

CHAPTER VI

High School and College

As a parent, I just assumed my children would attend grade and high school. It was also my hope that all would graduate from college. Not only did my daughters get a college education but I am providing financial assistance to my grandchildrens' college education. As you will see this was not the case in my time. Funding a high school and college education was a formidable undertaking but at times obtaining parental consent became even a greater impediment.

I GRADUATED from grade school in 1925 and by that time my older brothers and sisters had left home leaving me as the oldest one at home. None of my older brothers or sisters entered high school and to my knowledge none was encouraged to do so by my parents. You can imagine my parents reaction when I revealed that I would like to go to high school. With the older brothers gone, who would do the farm work? My campaign for a high school education picked up some support from my teacher, relatives and especially my oldest brother Ben who lived in Hays at the time. He offered to provide room and board for me if the folks consented. One of the arguments Ben used was the fact that my grades in school rated near the top of my class. Despite my campaign, the final decision came before the fall semester began at Hays high school and I was not a member of the freshman class. Needless to say, I was very disappointed.

Convinced that I had reached the end of the line with respect to my formal education, I took up my new occupation as sort of an executive vice-president of the farm. This was about the time Dad purchased a tractor and we had leased 160 acres from J.C. Penney, not far from the land Dad owned. With the new tractor and plow we were able to turn some of the leased ground that had returned to sod and plant it to wheat. A year later J.C. Penney wrote us a letter congratulating us for the good crop saying it was the largest crop produced on that quarter. My suggestion to Dad was that we summer-fallow the northeast corner of the home place by opening trenches with a lister in the fall to catch snow and rain and then to split the ridges the following year before planting. It was during this operation that I escaped serious injury or death. The lister was pulled by a team of four horses. One mare of the team had a colt following her. After a day or two the horses followed the ridges and very little guidance was necessary. I was sitting on the machine daydreaming when the rear wheel of the lister fell into a badger hole and I toppled off the machine. As I fell, a severe blow hit my head and blood started down my cheek under the straw hat. Apparently, the colt was following behind its mother and my fall frightened the colt, whereupon it fired at me with both hind hoofs. After walking the team back to the house, I was taken to a doctor in Hays for some repairs to my scalp.

To state that my parents were reluctant to send me to high school because the farm needed my help, was not the only reason. They were also afraid the city kids would lead me astray. Apparently they had second thoughts about denying me the privilege of a high school education. A year went by and the subject never came up but to my astonishment as fall approached they announced if I still wanted to attend high school they would not stand in the way. My reaction to this change of heart was one of mixed emotions. Bitterness because I had lost a year of schooling and had fallen one year behind classmates but it also presented an opportunity that I thought I would never get. I swallowed the bitterness and made preparations for high school. Ben and his lovely wife Katie remained firm in their offer and I had a place for board and room. Little did Katie know what she was getting into. There were more high school kids to come.

The fields had been prepared for planting before high school opened that fall but the rains did not come. Dad did not want to plant the wheat in dry ground for fear something would devour the kernels before they had a chance to sprout. The rains finally came but then it was time for school. After notifying the school, I spent the better part of two weeks planting wheat. We did not have a new large drill usually employed with tractors but we attached a hitch to the tractor and then trailed three horse-drawn drills behind the tractor. The weather turned cold after the rain and I was wearing a fur-lined coat to keep warm. The wheat got planted and I made my entrance at high school, albeit late.

Ben lived about four blocks from high school so the distance from my room to school was a short walk. One of my elective courses was typing and the school had just recently purchased new hand typewriters with left-hand carriage return levers. Unfortunately, by the time I arrived the new typewriters had been assigned so I took typing on one of the old right-hand carriage return lever machines. George Gatschet taught a number of commercial subjects including typing. I can still see his large sign on one of the walls, "Accuracy first, speed later." I must have taken it seriously according to my standing in the spring statewide typing contest. After the papers were graded Mr. Gatschet informed me that my copy was the only perfect typed copy turned in. That recognition was shattered, however, when the final grade at Manhattan discovered an error.

My stay with Ben and Katie was very pleasant and I doubt that I or my folks ever compensated them adequately for their generosity. When Friday night came the folks would arrive to take me to the farm for the week-end. On Sunday night a return trip was in order. One of the contributions the folks made consisted of food supplies from the farm. Katie made the best tuna casserole and her raisin-sour-cream pie was out of this world. One of Ben's neighbors taught latin in high school. He probably never forgave me for not taking latin. My commercial teacher, Mr. Gatschet, and I would attend daily mass in Hays. He was a graduate of the Weslyan at Salina and a convert to the catholic church. His brother worked for the Des Moines Register and flew a small plane. He flew to Hays one time and I had the pleasure of turning the crank to start the plane and then enjoyed a free airplane ride.

Growing up in a close-knit farm family is an experience that is most valuable and an opportunity many children never have. On the other hand, the isolation of the farm family from the public also created shyness and a reluctance to associate with strangers. Just getting a chance to go to town was a real treat. Sometimes I convinced the folks to take me to the dentist to have a tooth pulled when it probably could have been pulled at home. I enjoyed visiting with my classmates but I never got too chummy with them. For one thing, I worked some while going to school and that ruled out parties. We were warned not to take part in football because it was too dangerous. "After all, see what football did to Jim Finnegan."

Once the ice was broken, my younger brothers Cyril and Ray followed me into Hays high school. Why Katie ever put up with the three of us, I will never understand. It was in my second year in high school when I was thrust into some extracurricular activities. Maude McMindus was principal of the school and taught geometry. One day in class she placed a chair in front of the class and had some of us face the class to answer questions from the students. At the time it was most unusual and seemed out of place in a math class. Sometime later my name appeared on the main bulletin board, "Walter Schmidt see me—MM." Not many students get called to the principal's office. When I was made comfortable in the principal's office she said, "I suppose you were wondering why I asked you to face the students in geometry class the other day. It was to recruit members for our debate team and you have been selected." I told her, "Miss McMindus, that is one thing I cannot do. I freeze when I get in front of an audience." She replied, "Yes you can and one of these days you will thank me for making you do it." (I have many times.) From that day forward I was on the debate team. Later I was asked to enter a contest sponsored by the Building and Loan Associations and the subject was THRIFT. With the help of my debate coach, we prepared a dissertation on the value of thrift and practiced it religiously. One day I told my coach, "Since the contest is sponsored by the Building and Loan Associations, shouldn't we put in some plugs for the institutions?" My coach assured me the subject was THRIFT and not the importance of the Building and Loan Associations. Was he ever furious when the judges rated the speeches one, two and three. The one who spent most of the time praising the associations took first prize, the one that spent some of the time praising the associations took second prize and I took third prize and was awarded a twenty-five dollar gold piece. Another job I inherited was delivery of Lincoln's Gettysburg Address on Lincoln's birthday every year.

My best grades in all of my public education were in the commercial or math subjects. The lowest grade was is grammar. In one of the commercial classes Mr. Gatschet brought a clipping from the daily newspaper listing the successful candidates who passed the CPA examination. He told us about the importance of the CPA certificate and indicated that at one time he aspired to becoming a CPA but later elected to teach accounting. My grades in commercial subjects also got the attention of the principal. She asked me if my goal was to become a CPA in view of the grades in accounting, etc. I told her I had no idea what I wanted to become. In retrospect, she was a good judge of aptitudes. Many years later I did become a CPA.

The years went by as we spent the larger part of the year in school and the summer months on the farm. In both winter and summer I and my brothers worked at other jobs when possible to earn a little money. One year Cyril and I pitched wheat to a threshing machine for a neighbor and the going wage was thirty-five cents per hour. The days began about seven in the morning and lasted until dark at times. At night we slept on a quilt thrown on a wood floor in the haymow of the barn. An elevated stock tank was used for our showers. One semester my uncle Jake Miller offered me a job of painting his house and installing a chain-link fence around it. Jake was working at the Farmers Elevator at the time and I went there to discuss the job with him. After agreeing to accept the job it dawned on me that we did not discuss the rate of pay. Back to the elevator I went to resolve the oversight. We agreed that thirty-five cents an hour was reasonable pay. I apologized for taking so much of his time and Jake with a grin said, "By golly, that's business like."

My brother Ray told me that on his first day in class, Mr. Gatschet introduced him as the brother of the most handsome young man he ever taught. I think Gatschet went overboard in his introduction but he and I remained good friends over the years. It is my understanding that he taught a class at Fort Hays State College this year and that he is past ninety years of age. I also have a soft spot in my heart for my high school principal, Maude McMindus. After the high school faculty elected me and four other seniors to the National Honor Society and before the end of the senior year, she called me to the office for a little motherly advice. She urged me to set my goal high enough. She told me her goal was to be a high school principal and that her goal was not high enough. She urged me not to get married but to finish college first. When I told her it took money to attend college, she contacted president Lewis of the college to see if there might be an opening in his office.

In those days grades from one to eight were taught in elementary school and grades nine to twelve were taught in high school. I graduated from Hays high school in the spring of 1930 and enrolled at Fort Hays State College that fall. Like many of the state colleges at the time, Fort Hays was primarily a teachers college. So to begin with my major was education. People would ask me what my major was in college and when I told them, "BS in Education" the response invariably was a wide grin. I changed one of the subjects after attending one class session. As an elective I had selected astronomy and with Dr. Zinzer and his strange accent proclaiming that "Little Jupiter was about four million miles from earth" convinced me it was too far out for me. I finally selected sixteen hours of college work for my first semester.

Economically, there probably has never been a more difficult time to finance a college education. The stock market crash had taken place the year before, unemployment was reaching staggering proportions and the farmers were going bankrupt. I found work at a service station managed by a relative and worked fifty-five hours each week on evenings and week-ends. Sixteen hours of college work and fifty-five hours of service station work left very little time for meals or sleep. It was permissible for me to take my college books to the station and when business was slow I did study while on duty. In addition to the sixteen hours of credit courses, I was selected for the college debate team. In both high school and in college I was on the affirmative team. The subject for debate in high school was "Is the parliamentary form of government in Great Britain preferable to the congressional form of government of the United States?" In college the subject was "International free trade versus international tariff barriers." Fifty-five years later our elected representatives are debating the same subject. Elizabeth Reed was my debate partner in college and we won a number of debates. After one of our wins, I happened to hear someone talking to Dr. Reed in which he had some kind words for me but failed to mention Elizabeth. He evidently was unaware Elizabeth was Dr. Reed's daughter. Dr. Reed's response was, "Yes, he has a good voice." Elizabeth was a good debater and deserved recognition.

By the time I completed my first semester in college my brother Cyril had reached his third year in high school and my brother Ray was a freshman. In other words, there was no one at home but Gilbert and he had reached the ripe age of eleven. The folks were having a terrible time making ends meet and there was constant pressure from creditors. Payments could not be made to the Warren Mortgage Company holding a first mortgage on the farm and bills at retail stores in Hays were accumulating. The market value of the farm was dropping below the principal balance and the interest payable on the mortgage. Not only was I tired when I saw the folks at intervals but very vulnerable to the constant pleading and complaining by the folks. The work on the farm was getting too much for Dad and Mother. I finished the first semester with an average grade of B and started the second semester. The problems on the farm were mounting and so was the pressure for solutions. It is never pleasant to see people struggling to make a living but when it becomes your parents the pleas for help can become unbearable. The decision I made at this point in my life has disturbed me to this day. It appeared obvious to me that one of us students should lend a hand. Cyril had one year of high school left and Ray had just started. The rest of the story is history. I did not ask either to interrupt his high school education. Sometime during the second semester of college I approached the college and my employer in tears and told them I had decided to take a break in my education to help the folks. □

CHAPTER VII

Local CO-OP Manager

Little did I realize that accepting the manager's job of a bankrupt local oil cooperative would lead me to a lifetime of service to rural cooperatives. We speak of a farm depression today but economic problems of today cannot compare with the disaster of the thirties. The difficulties involved in trying to resurrect a defunct organization are enumerated in this chapter. It is also the period when I became a married man and started a family.

IT IS not in my nature to feel sorry for myself and I have told my daughters that the greatest disease of mankind is self-pity. But the truth is that the first two years after terminating my college education were the most difficult of my life. Never before and never since have I been so depressed. It just appeared that wherever I turned a mountain would rise before me. Farm prices dropped to their lowest in 1932. That was the year the market for a bushel of wheat reached twenty-two cents and a bushel of ear-corn reached nine cents. To this day I am not certain that dropping out of college actually helped anyone. Both of my younger brothers dropped out of high school and managing a farm without authority became a frustrating experience. Even the best farmers of the time were going bankrupt.

Here I was twenty-two years of age, not needed at home, no money and no job. Oh I picked some corn one year making about a dollar fifty cents per day plus room and board. That would hardly pay for replacement of the overalls and gloves. The thought occurred to me, why not try for a county teaching certificate? Teaching country grade school was not very lucrative but it beat picking corn. The examination for the certificate was given by the County Superintendent of Ellis County. Luckily, I passed the test and was awarded a county teaching certificate. The next problem was to find a school. The monthly salary for a country teacher was in the range of forty-five to sixty-five dollars per month. With some excellent reference letters from my high school principal and teachers, I started contacting country school boards for a job. Relatives are a good contact for such an undertaking. A first cousin was on the school board in a school district in Rush County and he offered to take me to see the two other board members. It was near the Memorial Day holiday so we visited the cemeteries while we were making the contacts. My cousin, Jake Herman and I stopped to see George Huber and Adam Weigand on our way. While at Mr. Weigand's place, Jake noticed Adam was building a new house not far from his large farm house. Jake asked, "Adam why in the world are you building a new house? You have that fine large house for you and Mrs. Weigand." Adam responded, "Jake, I have learned something, no house is large enough for two wives." The second house was for a son and his wife. After we had met the two other board members and were on our way to LaCrosse, Jake asked, "Why do you want to teach school? Why don't you apply for our manager's job? He made $150.00 during the summer months last year. We can hire a teacher for forty or fifty dollars per month."

Before the day was over, I agreed to meet with the Board of Directors of the Farmers Union Cooperative Oil and Gas Company of LaCrosse, Kansas. This was a rather new cooperative formed by farmers to provide savings in the cost of tractor fuel. It was a nonprofit organization with a five-member board of directors. There first manager was on a commission basis and developed a substantial sales volume but failed to collect cash for a lot of deliveries. When the storage tanks were empty and the bank account depleted, the manager, Paul Gibson, resigned to start his own oil company. Gasoline for tractor use was exempt from the state gasoline tax and was selling for less than ten cents per gallon. While the cooperative was out of business, Mr. Gibson was working overtime to purloin former customers of the cooperative. One of the inducements was a contract to supply the co-op member with gasoline for the entire summer at six cents per gallon. The Board of Directors met in early June of 1933 to interview applicants for manager of the cooperative. Actually, they were looking for a manager, tank-wagon delivery man, bookkeeper, collector, purchasing agent, organizer and public relations executive all in one individual. There were no employees to supervise at this point in time. It seems to me there were four or five applicants. One of the questions was, "Can you conduct the business of the cooperative on a cash basis?" To which I replied, "Frankly, I don't know but I am willing to try provided that cash basis means cash to the manager, the Board of Directors, as well as, each member of the cooperative." After the applicants were dismissed the Board of Directors voted and I was selected as their new manager. The understanding was that I would be responsible for any uncollected receivables. I was to furnish the truck for deliveries and the cooperative was to pay me on a commission basis.

The cooperative had a warehouse on main street in LaCrosse, Kansas, and some gasoline storage tanks located along the Missouri Pacific Railroad tracks. Our next door neighbor was the Farmers Union Cooperative Elevator and Mercantile Company, a nonprofit marketing cooperative organized before World War I, after the Capper Volstead Act was enacted granting income tax exemption to such producer cooperatives. My employer and I had one thing in common, both of us were broke. The accountant at the elevator cooperative offered to rent me a room in his house and I accepted. The local Chevrolet dealer extended me credit on the purchase of a used truck. The motor in the truck sounded good but there were no doors on the cab. With some help from friends, a delivery tank was installed on the truck chassis and proper licenses were obtained.

With the organization of local consumer cooperatives also came an effort to organize a cooperative wholesale for these locals. Farmland Industries, Inc. of North Kansas City, Missouri was organized by Howard Cowden about 1930 under the name of Consumers Cooperative Association. It started with a capital of about $3,000.00 and now represents one of the largest business organizations headquartered in Kansas City. The Board of Directors and I agreed quite early to become a member of Consumers Cooperative Association. As time passed I received considerable recognition from the wholesale for work performed at LaCrosse. Some of the sales programs developed at LaCrosse were mailed to other managers. I was invited to speak at an annual meeting in Kansas City regarding our sales program.

During the last year or two as manager of the oil cooperative, I was also hired to manage the elevator cooperative. In other words, I had to deal with two Boards of Directors. Even though my tenure with the grain cooperative was short, it proved to be very valuable experience later in my work. Besides remodeling the office to provide for a separate board room I designed new forms for the control of grain purchases. It seemed to me the forms in use did not provide the information necessary. In the purchasing of grain, it is very important to sell the grain on the same market used to purchase the grain. Using the new form, I knew how much grain to sell before the market closed for the day. One side of the form recorded total bushels received while the other side provided the disposition of the grain as open, stored or sold. I made one mistake. I should have submitted the form for copyrights. I saw the same form years later and the printed form was copyrighted.

Resurrecting a business that fell on its face after its original flight, proved to be a formidable task. Many if not most of the members had signed contracts with the former manager for their summer supply of gasoline. Gasoline prices had increased and the cooperative had to charge nine cents per gallon and demand cash payment. At my suggestion, a number of public meetings were arranged at district school houses where I and one or more of the directors would attempt to lure the members and others back to the fold. The farm economy was atrocious at the time. Not only had the price of farm products fallen dramatically, the rains had stopped falling, March winds were eroding the top soil and the grain elevators were empty. Some of our meetings would attract perhaps eight or ten farmers. Across the street from our headquarters, in the warehouse, was a small cream buying station. The buyer, Leo Helfrich, told one of my helpers years later that when he first saw the new co-op manager in overalls getting into his truck without doors he said to himself, "Now the co-op did it." Delivering gasoline in those days was hard work. Trucks were not equipped with power take-off pumps. Gasoline was drained out of the truck tank into five gallon buckets. A hand operated counter was mounted on the bucket compartment on the back of the truck. Many farm gasoline tanks were mounted on an old wagon. To fill the farmers' tanks the deliveryman was required to stand on a small barrel, stoop to fill the five gallon bucket and then stretch to pour the gasoline into a funnel inserted in the opening of the farm tank. Oil came in thirty gallon drums and had to be loaded and unloaded from barrel hangers on the side of the truck.

Business was terrible for the first five years but especially poor for 1933 and 1934. In 1935 our former manager was forced out of business by the state. He apparently failed to forward the state gasoline tax to the proper authorities in Topeka. His trucks were sold at sheriff's sale. Enforcement of the cash basis system of doing business added to my problems. Shortly after I accepted the job, one of my directors ordered a tank of gasoline and paid for it with his personal check. Several days later the bank returned the check marked "Insufficient Funds." I made a trip to his farm with the check and told him, "You were at the meeting when I accepted the job and by your vote you asked me to operate the cooperative on a cash basis. Surely you must know you did not pay for the gasoline. If you fail to redeem this check within a week, I will turn it over to the county sheriff." The check was redeemed and he never gave me another bad check.

A similar problem developed later but the customer in that case ignored my threat and let the time for redemption expire. I took the check to the sheriff and signed a warrant for his arrest. The sheriff notified the customer of actions taken and allowed some time for him to pay. When the time expired the sheriff called, "Has he paid? Shall I pick him up?" I told him he had not paid and to pick him up. One of my employees came to me later and said, "I just saw the sheriff's car go by and our customer was sitting in the back seat." The father of the character in question made settlement and he was released. One customer owned a hatchery and I accepted chicks in payment of the bill. I worked for the local cooperative for nine years and copies of my income tax returns indicate deductions were taken for bad debts. Along with the depression came the Roosevelt Administration and federal subsidies to support agriculture. During the thirties it was virtually impossible for farmers to accumulate enough cash to put out a crop. Payment for gasoline came by way of assigning the federal wheat allotment checks to us.

For the first few years in LaCrosse I rented a room and ate meals at a cafe. I had a nice room upstairs where the manager of a cleaning shop and the manager of the dime store also had rooms. We had the upstairs to ourselves. My monthly rent was $10.50. My breakfast of ham and eggs at the Mecca Cafe cost 35 cents. As I became involved in other activities my brothers came to LaCrosse and we rented a new bungalow for $25.00 per month. I persuaded my service station manager to leave his job in a neighboring town and work for us at $60.00 per month. It represented a $10.00 raise for him.

My work at LaCrosse brought me in close touch with the county agent for the farm program and with the county treasurer's office. Frank Pivonka, a solid democrat and a strong proponent of farmer cooperatives, was county treasurer at the time. He also had an attractive, intelligent daughter working for him. I learned that Louise had acquired a good record as a country school teacher for eight or nine years. When Frank Pivonka's term expired he purchased a local insurance agency and again I dealt with him as my truck insurance agent. By then Louise was working with the farm loan agency and was playing the organ for the church choir. When I was invited to join the choir I accepted. While Louise and I were friends for sometime, it was not until about 1938 that I marshalled enough courage to ask her for a date. In a small town, the newspaper editor loved to write stories about people known in town. In the thirties the best paid job in town was the postmaster job and it paid $125.00 per month. The cashier of the bank made $75.00. One nice Sunday afternoon the bank cashier and I drove to the country where we recognized some farm boys on horses and stopped to visit with them. The boys invited Adolph and I to ride the horses and we did. The next issue of the local paper had an article stating that the bank cashier and the co-op manager were coming to the office with pillows in hand following their experiences. Louise and I were scared the same thing might happen to us. As I was taking Louise home one evening, we noticed the aurora borealis (northern lights) and to get a better view we drove a short distance east of town toward the cemetery and parked. When I attempted to turn the car around, the rear wheel inadvertantly dropped over the edge of the narrow road and I had to get the local junk dealer out with his wrecker to retrieve us. The story never reached the newspaper. Louise and I will be forever grateful to Mr. Carbaugh. It was during this time that my aunt Elizabeth, stopped by my office and offered me some motherly advice. She asked, "Why don't you buy a house? Your rental payments would make the payment on a house." Hoping to get rid of her, I said, "It's really quite simple, I don't have the money for the down payment." That just opened the gate for her, "You don't need the money. I will loan it to you." I purchased my first two-bedroom bungalow, about six years old, hardwood floors, a fifty-foot lot and a two-car garage for $2,100.00 with $600.00 down.

Louise and I became engaged in the fall of 1938 and were married on May 22, 1939 in St. Michael Church in LaCrosse, Kansas. My brother Ray was best man and Loretta Moran was bridesmaid. My 1935 Chevrolet took us to the Missouri Ozarks for a short honeymoon. We continued to live in LaCrosse for another seven years and two of our three daughters were born at St. Anthony's hospital in Hays about twenty-five miles north of LaCrosse, Kansas. Paula Kay and Anita Louise were born in Kansas while Nancy Elizabeth was born in Missouri.

Farmland Industries, Inc. of North Kansas City, Missouri was not very large when the oil cooperative in LaCrosse became a member and went by the name of Consumers Cooperative Association (CCA). I recall that Howard Cowden, president and his sales manager stopped in LaCrosse and we had a nice chat in the shade of our warehouse building. As a member of CCA we were expected to attend the annual meeting. CCA's office and warehouse was at 1500 Iron street in North Kansas City and the annual meeting was held in one of the school auditoriums. A dinner for the members was provided by the ladies of a local church. I and several of my board members drove to Kansas City and it took us a day for the trip. Highway

U.S. 40 bisected every little town on the way and then took us through Fort Riley where we were delayed for most of an hour to let the cavalry pass. We made it from LaCrosse to Junction City by lunch time. I recall we stayed in a hotel in North Kansas City and I took a street car on Sunday morning to cross the Missouri river where I attended mass at a downtown catholic church. It was cold and drizzling when I entered the empty street car and my introductory remark was, "This beats walking." Without a smile on his face the driver answered, "Not a hell of a lot." He had not been on the job very long and obviously hated the job. □

CHAPTER VIII

Entrepreneur

During the thirties the unemployment rate stood at 25% and for many it became a battle for survival. In addition to managing the cooperative I provided employment for members of my family and friends by venturing into a number of small business activities. As you will see when you read this chapter, none was very successful.

DURING the depression and the dust years of the thirties it took very little capital to acquire a business. Many entrepreneurs were anxious to sell. Labor was no problem. Many of my relatives and friends were out of work. So while I was employed by the cooperative in LaCrosse I became involved in a number of small business enterprises.

Many of the farmers survived largely because much of their food was supplied by the farm and any other necessities had to wait until excess cream or eggs could be marketed. There were several cream buying stations in LaCrosse operated by creameries located in Salina. One of their agents told me his creamery was looking for a contract carrier to transport their cream accumulated along U.S. 40 from western Kansas to the creamery in Salina. After some negotiations we reached agreement and I started the W-S Truck Line located in LaCrosse, Kansas. Proper licenses had to be obtained from the Kansas Corporation Commission and a suitable truck purchased. The truck would start at the west end of the route after closing time and load cream until it arrived in Salina next morning. The volume would vary depending upon how much cream the farmers brought to market. Loading consisted of lifting ten gallon cans of cream by hand from the ground and stacking them on the truck. When the truck arrived at the creamery in Salina the driver had to unload the cream. Not only was it nighttime work but hard work.

I made some of the trips in emergencies but generally one of my employees was assigned that job. While insurance was necessary, there was always the fear that a relative or friend might fall asleep and meet with a serious accident. There were calls at night when something went wrong with the truck or a tire blew out. Like so many businesses during the thirties, it provided work for some people, the proceeds covered expenses but it was not a very profitable undertaking. There were some accidents but fortunately no one was hurt.

A common carrier route from Great Bend, Kansas, that interlined with the Santa Fe Railroad became available and the railroad contacted me to see if I was interested. This route picked up general freight at the Santa Fe depot in Great Bend and delivered it to merchants and others along Highway 4 to the west end of Kansas. Since I was in the trucking business and cream was not picked up each day, I decided to add this to our business. The Santa Fe people provided assistance with my tariff filings and necessary licenses. The truck driver in this case had to obtain receipts for deliveries and during the thirties collect C.O.D. shipments. If the merchant could not pay, W-S Truck Line had to store the freight until payment was forthcoming. In normal time this probably would have been profitable business but during the thirties there was very little freight moving west of Great Bend. The proceeds covered operating costs including labor but it was not a profitable operation.

My brother Henry was farming in western Kansas and his wheat crop did not make it through the winter. The ground was suitable for the planting of row crops but it took money for seed and gasoline. So he and I entered into a joint venture. I supplied the funds and he provided the machinery and the labor. Early in the summer it appeared like we might reap a good crop. When the crop was about eighteen inches tall and waving in the breeze, rains ceased and the hot winds of July and August finished our crop. As fall arrived, Henry contacted me and said, "I plan to put wheat in the field where the row crop failed. The ground is in good shape and needs no cultivating but I feel like I should include you as a partner." I told him to proceed as planned without me. This ended my farming career.

The cooperative oil company had leased a small plot of ground behind the building to a cream buyer. The buyer apparently wanted a location adjacent to Main Street and moved his portable building to a location he had purchased. My brother Cyril was not employed so we came up with the idea of replacing the cream station with one operated by Cyril. W-S Truck Line picked up some used lumber along its routes and the building was erected. Cyril obtained his license and started buying cream for one of the creameries in Salina. It was about as successful as my other ventures.

I read a book one time in which the author contended that what a person achieves in life depends a great deal upon what use one makes of the hours between seven and ten o'clock in the evening. It was his contention that providing for a living requires all but three hours of the day and any success beyond a bare living depends upon the use one makes of those three hours of every day. While I was involved in all those activities at LaCrosse, I made use of the time by enrolling in a Higher Accountancy Course conducted by the LaSalle Extension University of Chicago. When I mentioned my enrollment to a cashier at the local bank he grinned and said, "Oh, they sold you one of those." I have never regretted the time and money I spent to complete the course consisting of five text books and problems for each:

Business law
Cost accounting procedure
Accounting systems—their design and installation
Auditing procedure
Corporate organization

Most of my studying was done in my small office at the Farmers Union Cooperative Oil and Gas Company at night. There is no record of when I started the course but I must have finished it about the time Louise and I were married. My diploma is dated February 2, 1940.

Completion of this course and the experience I had in management undoubtedly changed the course of my life. Glenn Fox, finance director of Farmland (CCA) dropped by one day to see if I might be interested in a job with that company. They did not want to offend the local cooperative as they were a member of the wholesale. He had discussed it with Mr. Cowden and both agreed they had an opening for a man with my credentials. The job entailed extensive travel, something I had not done before. It was understood that I would not have to move. Hitler had started World War II, draft boards were busy and rationing was in force. Because of rationing and the distances involved they hoped I would not come home each week-end. The wholesale had about six auditors in the field who visited local cooperatives regularly and conducted audits. My job was to conduct audits and supervise the managed stations. Managed stations, was the term applied to local cooperatives in financial trouble needing management assistance. Louise and I discussed the new job offer versus our future in so many activities in LaCrosse. In view of the net earnings at LaCrosse, the guaranteed annual income was enticing. According to my income tax returns, I never made enough money in any year from 1933 through 1941 to have income subject to federal income tax. There were aspects of the new job that we did not like but we finally accepted it as offering a better future for us than the situation in LaCrosse. I served notice to my employer that I wished to resign at the end of 1941. My trucks, building and equipment were offered for sale. I notified Farmland (CCA) that I would be available shortly after January 1, 1942. The new job paid $2,400.00 per year plus travel expense. Mr Fox assured me there would be annual merit increases. □

CHAPTER IX

Farmland Industries Inc.

Farmland Industries Inc. is really a new name for a cooperative over fifty years old. The original name was Consumers Cooperative Association (CCA). Local supply cooperatives constitute the membership of Farmland Industries Inc. This chapter covers ten years of my life as an employee of Farmland in four different capacities: As a public accountant auditing local cooperatives; As supervisor of managed stations; As staff assistant to the Director of Manufacturing and as staff assistant to the Controller.

UNWINDING my involvement at LaCrosse took some time and I needed a better car for the new job. World War II was in progress, cars had been rationed and my 1935 chevrolet really was not equal to the task. One of my first assignments was to visit a grocery cooperative in Phillipsburg, Kansas. With my connections at Farmland (CCA) I was able to find work at the Phillipsburg Co-op Refinery for two of my employees at LaCrosse, Emery Bachand and Marcus Gottschalk. Emery was driving a 1941, four-door, Chevrolet with low mileage. While in Phillipsburg I told Emery of my car problem and he remarked, "I just might sell you mine if the price is right." After a bit of negotiating I purchased his car for eleven hundred dollars with a nominal down payment and twelve monthly installments interest free. Many consumer items were rationed including food, gasoline and tires. The fact that I was working for a petroleum wholesale and worked closely with the local service stations, was an advantage when it came to getting coupons for gasoline. By the same token, my traveling and eating out was an advantage when Louise had to stretch her food coupons. My daughter Paula Kay was about two years old when I started traveling so Louise and Paula would travel with me at times. I recall they were with me at Phillipsburg, Mankato and Norton, Kansas and at Scottsbluff, Nebraska. I'm certain there were other locations.

Farmland (CCA) grew at a rapid rate in the thirties under the skillful leadership of Howard Cowden. He was a tall impressive looking gentleman and an outstanding orator, especially when it came to promoting the consumer cooperative movement. Farmers for many years had been organizing agricultural producer cooperatives but this movement would allow all consumers to pool their purchasing power, including the farmer. At first Farmland (CCA) was nothing more than a broker, purchasing petroleum products wholesale and delivering them to local cooperatives for retail. Before I accepted the job with the wholesale, it had entered the refining business by organizing the Cooperative Refinery Association, a subsidiary, and building its first refinery at Phillipsburg, Kansas. Not long thereafter, it purchased a refinery in Scottsbluff, Nebraska and several retail outlets from Terry Carpenter.

In the fall of 1942, my new employer had an auditor by the name of Rudy Zurbuchen come through LaCrosse, Kansas, and take me with him to Sterling, Colorado, the location of one of the local cooperatives in trouble. As I recall it was in early September and one morning we found two inches of snow on the top of his car. We continued our tour to Trenton, Nebraska, and other locations. In each case we simply helped them close their books and compile financial statements. A board meeting was then called to discuss our finding with the board and to report on our stewardship under the management contract. The management contract was an emergency measure to help the local employees and boards of directors establish local control over operations and to operate the business successfully. In essentially all cases the local people were able to resurrect their organization. Sterling, Colorado, was one however, that went over the brink and was about to disappear under the waves. One of their largest creditors, other than CCA, had sued the cooperative and obtained a judgment against the property. When the property came up for Sheriff's Sale, CCA handed me a signed blank check and asked one of their attorneys and I to attend the sale and to recover the property. Sterling also survived and became a successful cooperative under local control in due time.

One of the jobs I inherited was to transfer ownership of retail outlets purchased in Nebraska when the refinery was acquired. In some cases there were no local cooperatives and new ones had to be organized. In other towns it was a matter of convincing the local cooperative to join Farmland (CCA) and to purchase the facilities from the wholesale. In some cases the retail outlet was simply closed and the property transferred to other locations. The records for the refinery and the retail outlets were maintained in the offices of the refinery in Scottsbluff, Nebraska. Farmland (CCA) had retained one of Terry Carpenter's supervisors to continue control of the books and records applicable to the retail outlets. Funds were deposited at the various local outlets and the supervisor was authorized to withdraw funds for payment of payroll, bills and transfers to Kansas City. With the limited audit staff, Farmland (CCA) had not found time to conduct an audit of these records for a period of about four years. This disturbed some of the officials in Kansas City, so I was assigned the task of auditing the books and records of the supervisor at Scottsbluff. He was a competent accountant and had many years of experience with the retail petroleum business. His penmanship was beautiful. Balancing his books and reconciling the controlling accounts with subsidiaries was progressing satisfactorily.

I was staying at one of the nicer hotels in Scottsbluff but the clerk and owner had one problem. She liked to imbibe in the spirits and celebrate with the soldiers stationed in the area, at all hours of the night. On one of my tours through Nebraska I failed to call for a room reservation. When I walked into the lobby she greeted me with, "Oh my God, why didn't you call? I don't have a room for you." After a long pause she offered, "Take this room. I never liked this guy anyway" as she walked across the lobby and started to clear the room of the tenants belongings. I have no idea how she settled with that man.

One of the problems I encountered in my audit of the records at Scottsbluff was when I tried to reconcile the accounts payable on the Scottsbluff books with the accounts receivable on the Kansas City books. My working area was a rather large conference room on the second floor where I had plenty room to spread books and records on tables. One day the Supervisor came upstairs and asked, "Walt, how are you getting along?" I told him "Fine with the exception of reconciling your inter-company account." I finally called the office in Kansas City and told them the account was out of balance by about $10,000.00 and that I was handicapped in my effort to pinpoint the difference because not all of the monthly statements from the Kansas City office were available. Mr. Fox took the matter up with Mr. Cowden and due to the amount involved they decided it was worth the effort to find the missing pieces. Kansas City put some typists to work and they copied all the statements from their files that were missing in Scottsbluff. Armed with the bundle of copies I returned to Scottsbluff. When I entered the Supervisor asked, "What are you carrying?" I told him, "Copies of the statements I couldn't find."

While I was busy tracing transmittals from Scottsbluff with the credits on the statements from Kansas City for the Sidney, Nebraska outlet, I discovered some discrepancies. One day the Supervisor came upstairs and inquired about my progress and I told him, "Everything is checking out fine with the exception of the Sidney account. You show some transmittals to KC for which they show no credits." I was not aware I had touched a sensitive nerve. The next day the Supervisor failed to come to work. The gossip around the office was someone took him to Cheyenne, Wyoming for a physical checkup at the Veteran's Hospital. A day or two later the phone rang and someone wanted to talk to the auditor. It was the Scottsbluff Chief of Police asking whether the Supervisor had been discharged. I explained that was not the case and we expected him to return after his visit to Cheyenne. The Chief finally said, "I am calling from his house. He is dead, apparently from a bullet wound." At dinner that evening my mind kept going back to my disclosure that transmittals to Kansas City that he claimed were not credited at KC. After dinner I pulled all the checks payable to Kansas City that apparently never reached KC. All of them were endorsed by a bank in Scottsbluff. Next day I called the bank cashier for an appointment to discuss the matter with him. He was very reluctant to explain how these checks found their way into his bank. He confided that the Supervisor and his family had accounts with the bank and he did not want the news media to get involved. I convinced him the Supervisor was bonded and if there was an embezzlelement, I would simply file a bond claim and the case would be closed. The cashier finally told me, "He would come to the bank during the noon hour and cash checks made payable to Consumers Cooperative Association claiming he needed the change for the co-op stores and service stations in the city." When I asked him about the size of the checks he said, "They were in round numbers perhaps several hundred to a thousand dollars each." He also informed me that the Supervisor played for money at a local lodge.

With this information, I went back to the office to do more searching. In discussing the case with my colleague Zurbuchen, who was in town on another matter, we pondered the possibility of getting the banker to sign a written statement to confirm his oral disclosure. We agreed it was worth a try. The next day I went back to the bank and presented a typed statement to confirm what he had told me the day before and the cashier signed it. The next question was, where did the company's internal control procedure break down and permit such an embezzlement? How did the Supervisor obtain a second signature on the check from the office manager in the refinery division because he normally inserted the check in the outgoing mail after signing? The answer became evident when I learned the office manager went to lunch at noon and the Supervisor did not. It gave the Supervisor the opportunity to retrieve the checks and take them to the bank, after they had the two signatures. In time the death was ruled a suicide. The banker told me later, "Never will I sign such a statement again." The matter was finally resolved by the bonding company paying Farmland (CCA) half of the claim and the bank paying the other half. It was settled out of court, the bonding company claiming that CCA's granting authority to the Supervisor to sign checks did not grant authority to cash the checks.

The first five years of my employment with CCA was devoted to audits and management of local cooperatives and any other job that might be assigned to me. Even the selling of corporate debentures was not off limits. Life on the road was primitive compared with travel in the eighties. Most of the cooperatives were in small rural towns and I was lucky to find a hotel or motel in the town. Room with bath was the exception rather than the rule. Lighting in the rooms often consisted of a single bulb dropped from the ceiling with a string attached to the switch. Some of the hotels had two or three floors with a public bath on each floor. Heating was supplied by water or steam radiators that worked part of the time. Sometimes I had to find a room in a neighboring town and commute to the job. A lowered window on the car or an open window at the motel or hotel constituted the airconditioning available. Some of the larger cooperatives had an office. But in the managed stations and the smaller cooperatives I was lucky to find a desk or table to work on.

World War II made the first five years difficult. Young men qualified for the management of a local cooperative were being drafted for military service. Available help had to be trained for every segment of the business. I and a group from LaCrosse, Kansas, were sent to Leavenworth, Kansas, where I received a physical checkup, learned to make my bed, ate some scrambled eggs and when I returned to La Crosse the local draft board had reclassified me. While in LaCrosse, I received a phone call from brother Ray in Chicago. It was shortly after Japan invaded Wake Island. He told me he had enlisted in the Marine Corps and was on his way to San Diego. I told him, "You will be too late to help the Marines on Wake Island." He was shipped to the Pacific theatre of the war and fought the Japanese starting with Guadalcanal, on to Tarawa and all the way back to the Japanese surrender. Ray was single when he enlisted but both Cyril and Gilbert were drafted for service when each was married and had children. Gilbert was at the front in southern Europe and received the purple heart. Cyril spent his hitch in the army at Fort Riley, Kansas, but was transferred to San Diego, California as part of the army of occupation to be sent to Japan. The troops were being loaded in alphabetical order and reached their quota with the name Schafer. This was how close Cyril came to a stretch in Japan after surrender as a member of the occupation army. Gilbert got a break because he understood and could speak some German. When the officers discovered Gilbert understood the German dialect of Bavaria they made him an interpreter. He tells of being sent ahead of the troops to procure housing for the night. He knocked on the door of a large house and when the lady of the house appeared, Gilbert in his best German conveyed the army's request for a night's lodging. She looked at him and said, "You're American aren't you? Why don't you speak English?" Gilbert admits, "She spoke better English than I."

The war was not the only cloud hanging over our family. Mother was suffering from gall stone attacks and goiter problems. Through her German newspaper "Der Landtsman" she was taking some medicine prescribed by the paper but her condition deteriorated. Surgery was suggested but mother said, "What good will it do? Mrs. Weber has had two operations but the stones return." On July 24, 1943 mother passed away at St. Anthony's Hospital in Hays, Kansas at the age of sixty-one years.

By 1946 I had been on the road for four years, my salary had been increased to about $3,600.00 per year and we had two small girls at home. Anita was a little over a year old. There may be accountants who like constant travel but it is hard to believe there are many with small children at home. I had notified Mr. Fox that I was interested in remaining with Farmland (CCA) but in a capacity that did not require constant travel. One day I received a call from Mr. Fox while I was working at Sterling, Colorado. He advised that Mr. Cowden would be at the Albany Hotel in Denver on Saturday and would like to talk to me about my request. I agreed to meet him at the time suggested.

Howard A. Cowden was relaxing in his room with his feet propped up enjoying his favorite cigar when I made my entrance. After a short visit he brought up the subject by saying, "The manufacturing division is getting too large for Rudy Zurbuchen to manage without some help. I have talked to him and suggested that he hire an assistant to help him. Rudy is a very capable man but the division includes the paint plant, printing plant, vegetable cannery, fertilizer plant, alfalfa mill, soybean mill, feed mill and the lumber mill and he simply needs a larger staff. Despite my urging for sometime, Rudy has done nothing about it. If you are interested in a staff position with Rudy, I will contact him on my return to Kansas City and direct him to get in touch with you." I told Mr. Cowden that I was interested and would look forward to hearing from Mr. Zurbuchen.

It didn't take long for the wheels to start turning. Zurbuchen and I had been good friends as colleagues on the auditing staff so he knew my qualifications. I agreed some travel to manufacturing plants was tolerable. He agreed I would have an office at headquarters with a secretary. I agreed to move to the Kansas City metropolitan area and he agreed to pay the moving expense. A starting salary of $4,800.00, vacation and sick leave pay, plus retirement pension benefits were acceptable on my part. One of my audit partners by the name of Dana Flanigan reported a house for sale in his neighborhood in Independence, Missouri. We looked at the house and finally bought it. A trucking firm from Hays, Kansas, picked up our belonging in LaCrosse, Kansas and delivered them to 2901 Forest. Dr. baker in LaCrosse learned that we were moving and was interested in buying our house for a doctor on his staff. I priced the house at $4,500.00 and it sold for cash.

Our new home was a two-bedroom colonial frame house about five years old. It had a full basement and a stairway but the upstairs was not finished. It was part of a development containing perhaps forty or fifty homes. We purchased the place May 1, 1946 and paid $7,979.00 for it. Dana Flanigan lived in the same area a few houses closer to Chrysler Avenue. When I was in town, he and I would take turns sharing rides to the office located at 318 East 10th Street, Kansas City, Missouri. My office was on the sixth floor next to Mr. Zurbuchen's office. Our house was on a pie-shaped lot with considerable front footage but not so much space to the back. Louise called the area "Chrysler Gulch" because it was near Rock Creek and in the rainy season our basement would spring leaks. We were members of St. Mary's Catholic Church of Independence, Missouri, and Paula Kay rode a bus to St. Mary's Parochial School. The church was about a stone's throw from President Truman's home.

Mr. Cowden was correct in that expansion into manufacturing required additional personnel not only for the manufacturing plants but also at headquarters for management and accounting. A real need existed for better liaison between plants and headquarters. The monthly operating reports for the manufacturing plants fluctuated from month to month due to improper accounting. This was the first area I tackled and it was a sensitive area because cost accounting was a department of the Control Division and I was a staff assistant under the Manufacturing Division. The accounting changes had to be made diplomatically with the cost accounting department and the accounting sections in the manufacturing plants. In time I was able to install new charts of accounts in the field and at headquarters to correctly value goods in process for inventory purposes. Inventories of raw materials and finished goods only were causing operating reports to distort operating results. When I first started working for CCA, a man by the name of Gillespy was Controller. Mr. Fox, Finance Director, told me, "Walt, I just as well tell you because you will find it out sooner or later, CCA has two sets of records, one is on paper and the other is in Mr. Gillespy's head."

One day I was approached regarding the risk Farmland (CCA) was taking since it had purchased 500,000 bushels of soybeans and placed them in storage at its mill in Coffeyville, Kansas, without hedging the cost on the futures commodity markets. Someone asked, "Walt, you have had grain experience, what happens if the price of beans falls one dollar per bushel during the next year's operation?" My reply was, "The answer is simple. CCA could lose $500,000." The soybeans were processed through the mill and two finished products were made available for use in the feed operation and for sale on the market. One was soybean meal used as cattle feed and the other was soybean oil used in food products such as oleomargarine. The feed department supervisor contended hedging of the cost of beans would not work because the price of soybean meal did not parallel the cost of the beans.

After preparing some graphs and charts I convinced both him and Mr. Zurbuchen that by combining the sales price of the meal and the oil, it did rise and fall with the market of the beans. Once that was agreed upon, I was asked to prepare my recommendation for the hedging of beans at the plant. My report in writing covered several pages with some pertinent suggestions. Since Farmland (CCA) did not have a seat on the Chicago Board of Trade but the Farmers Jobbing Association in Kansas City did, I suggested that CCA channel its transactions through the Jobbing Association. The Jobbing Association was to sell the beans at prevailing prices to be delivered at a uniform rate over the next twelve months. Each month as the beans were processed into meal and oil the Jobbing Association was to purchase the monthly contract at prevailing prices. This means if the price of beans increases over the year, the price of meal and oil will increase, the soybean mill will make a profit larger than expected but when the purchase of the futures contract is made at a higher price the profit will be normal. In other words, the gain at the mill will be tempered by the loss on the futures contract. If the price of beans falls over the year, the mill will lose money but the futures contracts will be purchased at a profit.

My plan was approved and the Jobbing Association sold the beans on futures contracts and was authorized to purchase the contracts on a uniform basis as the beans were processed. The Supervisor of the feed department was to advise the Jobbing Association regarding purchases.

The price of the beans started to increase in price resulting in a favorable margin at the mill which the Supervisor of the feed department did not want to tarnish by purchasing the futures contract. Instead of complying with rules of the game and purchasing contracts as the beans were processed, he delayed the purchase for a lower market. The beans were being processed and there was no way for him to deliver the beans on his sales contracts so he was forced to purchase futures contracts at ever higher and higher prices. Losses started to appear on the operating reports so the Supervisor contacted Mr. Zurbuchen for purchase advice. With a proper hedge the purchase is automatic at the time the beans are processed. To make a long story short, Farmland (CCA) did not lose $500,000 but it did lose about $100,000 in its soybean operation that year. At the annual meeting of the members, the loss was blamed on the manipulations of the soybean futures market by some Chinese. I told Louise, I knew who was responsible for the loss and he certainly was not Chinese.

Years later the Farm Credit Administration assigned a team of auditors to prepare a financial forecast before approving a substantial loan for a new nitrogen plant to be constructed at Lawrence, Kansas. One of the auditors had experience with grain and hedging. He came to my desk one day and remarked, "They tell me you are familiar with the soybean mill operations. I made an analysis of the soybean operation and it appears to me the loss was due primarily to improper hedging. Am I in the ball park with my figures?" I didn't tell him but I had prepared a similar analysis and our figures were within a thousand dollars more or less.

While I visited essentially all the plants, management of the plant remained with the local manager subject to review by Mr. Zurbuchen. My position in the manufacturing division was staff assistant to the director of the division. One time I was asked to conduct some research regarding the relocation of the alfalfa plant from Pond Creek, Oklahoma to Holly, Colorado. Such items as utility costs, availability of acreage, local wage scales, property taxes etc. were investigated. The plant was finally located in Longmont, Colorado.

Farmland (CCA) and other regional cooperatives concluded they needed a nitrogen plant to supplement their fertilizer program and after considerable research elected to build their plant at Lawrence, Kansas. The Farm Credit Administration, through the Wichita Bank for Cooperatives, agreed to finance the major portion of the cost but we had to raise several million dollars of additional capital to qualify for the loan. I was asked to join a sales force including fieldmen, field auditors and a number of people from the Kansas City office to conduct a finance campaign to sell five and one-half percent corporate debentures. Funds were sought in nine states in which Farmland (CCA) had members. The territory assigned to me was the area south of Lawrence, Kansas.

After making a few contacts at banks and other institutions, I ventured down the country road to sell securities. Farmers had their own economic problems and many had limited funds available even if they approved construction of the plant. Several days passed and my sales record was nothing to write home about. I learned many of the rural residents south of Lawrence were Mennonites. One day I was making a sales pitch to a middle-aged farmer, when he commented, "You need to contact the old man on top of the hill. These people undertake nothing without discussing it with him." I took his advice and paid the elderly couple a visit. They were very friendly and invited me to the house for a cup of coffee. The lady of the house wore a little hood over her hair. The house was immaculate. I thanked them for their hospitality, left some literature for them to peruse and told them I would drop in later. On my second stop there were questions and discussions but still no commitment. On a third visit the elderly couple had reached a decision and invested $5,000.00. Within a relatively short time, my sales of debentures reached $100,000.00 and was top on the list of solicitors in the field.

On a routine visit to the lumber mill in Swisshome, Oregon, Mr. Zurbuchen and I spent about two weeks at the mill in December and never saw the sun. It rained constantly. Zurbuchen worked on management and marketing problems while I distributed pay checks for our payroll department to assist our internal auditing staff. Swisshome is near the city of Florence along the Pacific Ocean west of Eugene, Oregon. Farmland (CCA) had a substantial investment in the mill, timber and a housing development for company employees. The mill was in good condition using a double-cut band saw accommodating logs thirty-two feet long. The terrain around the mill is mountainous and logs were trucked to the mill pond. At that time we had no drying facilities at the mill. About twenty-five houses were built on the hillside north of the mill and were painted white with green roofs. The sign at the gate read, "Co-op Village." While it was December of the year, climbing roses were blooming near the office and the water lines to the homes were lying on top of the ground. The heavy rains at that time of year and the moderate temperatures result from the Japanese warm water current that touches Oregon.

A young couple from the Kansas City office went to Swisshome and remained there for a number of years. When they asked to be returned to the office in Kansas City, Mr. Rasmussen, the mill manager, tried to convince Mr. Zurbuchen that I would make a good replacement but I declined the offer. As a result Mr. Rasmussen brought more of his family into the operations. His daughter started working in the office and one of his sons was working in the mill. The lumber department was not very successful and one day he and his family walked out. To keep the mill functioning it was necessary for Mr. Zurbuchen and I to pack our bags and fly to Swisshome. He took over the management of the mill while I took over the office responsibilities. Since there was no motel in Swisshome, the frame office building was equipped with modern conveniences and space for two or more standard beds in the unfinished attic became our motel. At the time I probably weighed about 150 lbs. while Mr. Zurbuchen would check in at better than 200 lbs. There were two beds with bedding including two blankets on each bed. Rudy wanted the windows open and when I told him "I was cold last night," Rudy responded, "It was just fine. I enjoy the fresh air." The following night Rudy went to bed first and was fast asleep when I made it upstairs. I quietly peeled one blanket off his bed and added it to the two on my bed. Next morning Rudy arose first and said, "Perhaps we should close one of those windows, I was cold last night also." It took some time before he noticed that he had only one blanket and retorted, "No dang wonder, you stole one of my blankets."

A trip to the lumber mill was a trip to another world. For one born and reared on the plains of Kansas; who worked for rural people most of his life, it was breath-taking. Douglas fir trees eighty to one hundred feet tall and trucks bringing logs off the mountain side measuring as much as five or six feet in diameter. The Siuslaw River was just roaring by the mill and the rain kept falling but the men in the plant with rain coats and wide-brim hats just kept sawing wood. Trucks would dump the logs into the pond and the employee in charge of the pond would ride and guide the logs to the saw mill. One of the descriptions that baffled the KC payroll department was the position of "whistle punk." We had to explain that communication between the man operating the cable machine near the spar pole and the tree fallers and buckers who fastened the tongs to the logs was extremely important from the safety standpoint. The whistle punk simply stood guard over the fallers and buckers. When the log was securely fastened to the cables and the men were clear, he would blow his whistle telling the man operating the machine to pull the log hanging from the cable across the valley. Some of the accountants in KC wanted to apply for the job since it paid more than some of them were earning. In the vernacular of the lumber industry, the cable machine is called a donkey. One of the accountants in KC wanted to know, "Does CCA have donkeys at the mill?" For a while we told him, "Of course. Some of them have been running around the mountains so long their legs are shorter on the hill-side."

Keeping employees for the mill was another problem. Many of the workers were either single or separated from their families and moved from mill to mill to break the monotony or to see which mill provided the best food and housing. One of the tree fallers was a young college graduate and I asked him, "What brings a young college graduate into the woods to fall timber?" His answer came in the form of another question, "Where else can I make as much money per day?" In addition to the office and the co-op village the hillside also included a meeting hall and a dining hall housed in a converted barn. Cooking and dining took place on the first floor. Meetings and entertaining took place in the haymow. A young man entertained the residents with a projector but since he was not licensed, admission consisted of a voluntary contribution to his gallon bucket at the door. He told me, "My bucket is getting empty. They just don't come to the better pictures. I'll bring John Wayne back and fill the bucket again." As mentioned before, the fringe benefits of good housing and food were important. Cognizant of that fact, Farmland (CCA) employed a married couple who had cooked at the Multnoma Hotel in Portland, Oregon. We fed our harvest crews well on the farm but I have never seen a crew better fed than a lumber mill crew. The mill manager told me he discovered the man had a drinking problem but it did not seem to affect his cooking. As the manager was sitting in his office one afternoon he heard what sounded like bullets whistling by the office. Investigation disclosed the chief cook had been imbibing and was shooting cock-roaches with a forty-five automatic.

Some of the manufacturing endeavors Farmland (CCA) undertook were successful because they filled an existing demand. This was not the case of either the lumber mill or the vegetable cannery. As a result neither proved profitable. Most members were local cooperatives marketing farm products or purchasing farm supplies. Some of the members did both. Very few of the members had a successful lumberyard or a competitive grocery store. As a result production from those plants had to be marketed through competitors and frequently at discounted prices.

The cannery was purchased from the Otoe Food Company of Falls City, Nebraska and was located at Scottsbluff, Nebraska. The North Platte Valley had ditch irrigation along the river and produced such crops as beans, sugar beets, peas and sweet corn. The main products for our cannery were sweet peas and sweet corn. The peas were canned by sizes and the corn was either cream style or whole kernel. One year we canned over

92,000 cases of corn. The plant operated for about ninety days each summer. Canning ceased with the first killing-frost. Pack was reduced some years by hail storms. On one of my first visits to the cannery, Mr. Zurbuchen and the plant superintendent were giving me a tour through the plant and as we passed the superintendent's little make-shift office, Rudy called my attention to the pin-up girls decorating the walls of the little office. "Tell Walt why you leave those pin-up girls on the walls." The superintendent replied, "Before I hire a man, I bring him by my office and if he fails to notice those girls, he's too darn old to work here."

At harvest time, peas and corn were harvested and brought to the plant by truck. The plant had two processing lines. One was for peas and the other line was for corn. We operated in three shifts of eight hours each and employed about 150 people, mostly women. Conveyor belts moved the product to work stations where women in white gloves prepared the product for canning and cooking. Machines sorted the peas and other machines would cut the corn either whole kernel or cream style. Filling machines measured the proper amount of product into the cans and firmly fixed a lid on the can. Batch code numbers were stamped into each lid. Large metal baskets carried the canned goods to pressure cookers in an adjoining room for final processing. From there the baskets were rolled on overhead rails to the yard where water sprays cooled the canned goods before they are packed in cases and stored in the warehouse. There was another conveyor system that brought empty cans from a railroad car or from storage to the filling machine. All food products were inspected and graded by a representative of the U.S. Department of Agriculture. The waste material was conveyed from the lines by water flumes and stacked to be sold as cattle feed. During World War II the government was in the market for dried or powdered potatoes. The manager of the plant suggested that Farmland (CCA) add a potato line in the plant that would dice and dehydrate the potatoes for sale to the government in thirty gallon barrels. He had a friend on the west coast, where he formerly worked, that had such equipment. The equipment was purchased and carloads of dehydrated potatoes were shipped from the plant in Scottsbluff, Nebraska.

I got involved in plant management one summer quite by accident. While we had a plant superintendent, with three shifts there were times when the manager had to relieve the superintendent. During one of my visits to the plant, a problem arose at the plant, serious enough to shut the plant down and the manager could not be found. I called Mr. Zurbuchen and alerted him to our problem and without hesitation he issued this order, "Fire him and take charge of the plant." Something should have been corrected with the waste flumes but was neglected by management and as a result the flumes became clogged and the plant had to shut down. With some help I corrected the problem and the plant went back into operation. While managing the cannery I interviewed applicants for employment. There was no Indian reservation close to Scottsbluff but one day a huge Indian came in to apply for a job. I had no intention of hiring him but I took his application. "What's your name?" I asked. "Left-hand Bull" came the reply. Another evening I was in the office and the plant was operating flawlessly when someone came calling me to the plant. Unbeknown to me, the man in charge of the cream style corn filling machine had brought a bottle of whisky to work. The contents apparently took effect and instead of the cream style corn dropping into the can it was spilling over the reservoir on the machine and spreading over the concrete floor around the machine. The starch in the corn made the floor very slick and when I arrived my man was lying on the floor in a sea of corn unable to get up. Thanks to a dedicated employee we pursuaded a day-shift operator to come in and take his place. Another problem arose when our Safety Director paid a visit to the plant. Part of his equipment was a little manikin he used to explain artificial, pulmonary resuscitation. Larry put his little manikin in the janitor's supply room before meeting with the employees. We had an elderly gentleman as a security guard and he kept his gun in the same cabinet. He came to work one evening and upon opening the door the manikin fell out at him. It scared him so he came to me immediately and offered his resignation. I finally convinced him this was no employee prank and he remained on the payroll.

The former manager of the plant released a written blast at Mr. Cowden, president of Farmland (CCA), in which he called Rudy and me "Stooges" and told Mr. Cowden, "The acquisition of the cannery itself was a masterpiece of stupidity."

Consumers Cooperative Association and Farmland Industries Inc. are one and the same nonprofit cooperative corporation founded by Howard A. Cowden. More significantly than the name change, has been the change in economic philosophy of the company with the passing of Cowden. Under his leadership, company expansion was financed through deferred patronage dividends and sales of debentures to members. A strong equity position was important. With his passing, the company grew rapidly but so did debt. As I write my autobiography the company is shedding subsidiary corporations and liquidating assets to stay afloat.

Dick Bland, controller of CCA and its subsidiary CRA, was guarding his territory. It was his contention that I, holding a public accounting certificate in the State of Missouri, should be under his control and not under the Director of Manufacturing. He finally convinced Mr. Cowden and my position changed from staff assistant to the Director of Manufacturing to staff assistant to the Controller. In this capacity I still worked closely with the cost accounting department but I also worked with general accounting, the budget department and the SEC Registration Section. It was during this time that I began giving serious consideration to public accounting as a professional career. There was a missing link however. I had no intention of leaving CCA for an accounting career without a CPA certificate.

Since I was a licensed Public Accountant, I qualified to sit for the CPA examination in the State of Missouri. With my new job in the Control Division I was home most of the time so I dusted off my accounting courses and applied for the International Accounting Society's CPA coaching Course, consisting of about 400 problems taken largely from past CPA examinations. At the same time, I enrolled in some courses offered by the Kansas University Extension Division. It was my intention to devote at least three years of intensive study to pass the examination before entertaining any other goals. I worked every one of the four hundred assignments and reread every text of the LaSalle Accounting Course. In May of 1949 I sat for my first CPA examination in Kansas City. The results of the examination were announced about ninety days later and I had passed the subject of law. The examination covers auditing, law, theory and accounting problems. At that time passing only in law meant I had to retake the examination in the four subjects. Once I started reviewing for a CPA examination I never interrupted study waiting for the results to be announced. I sat for the examination again in May of 1950. When the results were announced about ninety days later I had failed all subjects. I sat for the examination again in May 1951. About ninety days later my telephone rang and Louise was on the phone, "You have a letter from the Missouri Board of Accountancy. It's thicker than the other announcements. Shall I open it?" After she had opened the envelope she said, "Hang on to your chair. You passed all four subjects."

While I was taking the examination, I had lunch in CCA's cafeteria with the company attorney and he made the remark, "If you are successful and receive the CPA certificate, there certainly should be a spot for you right here. Do you realize this large company does not have a CPA on its payroll?" For that reason I did not expect big things from my employer. Furthermore, my ambition was to enter public accounting. Howard Cowden, President, was in Copenhagen Denmark at the time and mailed this short note, "Congratulations on your CPA." Sometime later my monthly pay check was increased by twenty-five dollars.

Bookkeepers or accountants have been a part of every society for centuries. As the industrialized nations depended more and more on capital furnished by the public to finance industry, the need for specially trained competent and independent accountants who could examine financial statements issued by industry and report on their findings became essential. Certified Public Accountants were first licensed in 1896 in New York. Hundreds of thousands of CPAs are licensed by the fifty states at this time. The American Institute of CPAs is the national professional society of CPAs. There are state societies of CPAs in each of the states. In order to become a CPA, one must satisfy education and experience requirements specified by state law, and pass a difficult two-and-a-half-day written examination. The examination is prepared and graded by the American Institute and is uniform in all states, territories and the District of Columbia. Virtually all successful CPA candidates at the present time are college graduates.

The 1951 examination was taken by fifty-five candidates in Kansas City and five of us passed. About two hundred and fifty took the examination statewide and about ten percent passed. Very few candidates pass the examination on the first attempt. It took me three tries and that is about average for all successful candidates. In 1985 about 1,300 applicants took the examination in Missouri and about twenty percent received certificates. Licensing of CPAs is the prerogative of the states in accordance with the statutes of the state. Boards of Accountancy of each state, appointed by the governor of the state, enforce the rules and regulations pertaining to CPAs. At the present time I practice under license granted by Missouri and hold membership in the American Institute of CPAs and the Missouri Society of CPAs. CPA certificates were issued to me over the years by the states of Kansas, Wyoming, Nebraska, Iowa and Louisiana. It was not necessary for me to take the examination in those. These states recriprocate with the state of Missouri in the issuance of the CPA certificate. While I was struggling with the CPA examination, I had lunch with a member of the Missouri Board of Accountancy and I asked him, "Am I just spinning my wheels or do I have a chance to pass the examination?" To which he replied, "Certainly, you have a chance to pass the examination. Only a few hot shots pass it the first setting. I can't speak for other states, but as you know, the American Institute grades the papers by number and arranges the two hundred and fifty papers in order by grades before returning them to us for review for CPA certificates. We do not review all the papers but select perhaps one fourth of the top grades and from that sample award the certificates."

Public accounting and auditing came into being in the 1850s in Great Britain under the name Chartered Accountants and was practiced as such in America as industry developed. Professions in fields such as law, medicine, engineering and architecture were established long before the public accounting profession emerged. While the professions are subject to statutes of the states they are actually licensed and regulated by their peers serving on the state boards appointed by the governor. People have asked me, "Why is it so difficult to penetrate your profession?" There has never been any doubt in my mind that when the state boards admit new members to their profession they are keenly aware of supply and demand. It cannot be denied that each profession has absolute control over the supply of licensed professionals. The medical and the public accounting professions have admitted the smallest percentage of applicants to the profession. The passing percentage of successful CPA candidates about doubled in the last thirty years. The legal profession has consistently admitted most of its applicants to the bar over the years. The legal profession must have concluded that members unable to find a position in industry or in private practice have the alternative of entering the political arena. CPAs will tell you, and rightly so, we set our standards high to protect the quality of service to the public. Many industries employ CPAs for their finance, accounting or treasury departments.

A story on accountants in a 1932 issue of Fortune magazine put it this way:

> *Towards the turn of the century there sprang up in this country a new profession. Already established abroad, it followed English investments to the U.S., watched over them, took root. Today, it is no overstatement to say that there are preeminently three professions upon whose ethics as well as upon whose skill modern society depends; law, medicine and Certified Public Accounting. Yet this third profession,*

> *which is no heritage but a creation of our necessity, is so little known that certainly 100,000 of our free, white, educated American contemporaries will die in A.D. 1932 without ever having the vaguest of what manner of men Certified Public Accountants are. They walk in the shadow of virtual anonymity. So discreet are they that at times it seems as if their aim were to become a disembodied function, almost without proper name. Yet upon expert opinion of these abstract beings, who pit their judgment against the unbelievably subtle economic forces of this generation, the financial structure of our greatest industries is founded.*

After the 1907 panic, the Federal Reserve Board was established in 1913, followed by the Federal Trade Commission and the enactment of the first federal income tax law. The stock market crash of 1929 resulted in the creation of the Securities and Exchange Commission (SEC) which required audits of corporate financial reports and did more than anything else to foster the enormous growth of public accounting. Auditing remains the foundation of the large international CPA firms. Eight of the largest CPA firms in the United States audit ninety percent of all the corporations listed on the New York Stock Exchange. Big business needs large auditing firms with offices located in many nations. Audit fees for a single client can vary from tens of thousands to millions of dollars for the annual audit. Large CPA firms rely largely on personnel recruited from college graduates.

Litigation has become a major problem for CPA firms auditing public corporations. Investors and creditors of bankrupt public corporations, subject to tremendous losses, retaliate by filing law suits against the directors of the corporations and the CPA firms whose audit reports are part of the annual statement to SEC and part of the prospectus for the sale of stock or debentures. The claim is that the CPA firm knew, or should have known, the corporation was in financial difficulty and failed to disclose that fact in their reports to the public. Large court awards have been assessed against CPA firms and premiums for professional liability insurance have skyrocketed. Some insurance companies no longer insure against professional malpractice. The investing public and the creditors are unreasonable when they expect the CPA firm to forsee and disclose every possible contingency that could befall the corporation. Or for the doctor to forsee and disclose every possible condition that could affect the health of his patient. Thank goodness, laws are being drafted in Washington and enacted in some states to deal with the ridiculous awards to plaintiffs, in not only the malpractice suits, but in other liability suits.

The uniform CPA examination may be a good test of the candidates knowledge of the necessary accounting and auditing skills but a successful CPA firm needs competent CPAs with unquestionable character and integrity. A CPA that sells his opinion, should not be a CPA.

With the large audit fees, there is always the monetary pressure by the underwriters who want to sell the corporate stocks or bonds and the public corporation that needs the capital from such sales. The underwriter may tell the CPA firm, "Unless your firm grants the corporation a clean bill of health (an unqualified opinion) we will not be able to handle the sale of your client's stock or bond issue." The client may say, "We had no trouble with our previous CPA firm and if you qualify your opinion, it will be your last audit of this corporation." The CPA firm is in the uncomfortable position of qualifying its report and risking the loss of a valuable client or granting an unqualified opinion and risk litigations should the corporation encounter financial problems. It has been said there are two sides to every question. There probably have been CPAs who found questionable conditions or practices during their audit and failed to report them for fear of losing a client. It is hard for me to believe that such a CPA would remain on the payroll of a successful CPA firm for long or remain in private practice for long.

While a large percentage of CPAs in the United States are employed by the large auditing firms and are assigned to conduct audits almost exclusively, most of the CPAs are involved in smaller auditing firms that conduct audits for hundreds of thousands of smaller corporations, partnerships and individuals engaged in thousands of businesses and other activities. Thousands of CPAs work for federal and state governments. Auditing is not a major activity in many small partnerships or private practices. Income tax work, unaudited financial statements, bookkeeping assistance, management advisory service, and personal financial assistance constitute the major work of these firms and individuals.

I doubt if anyone knows how many CPAs there are in the United States. The American Institute of CPAs, in one of its publications dated May 1986, does provide us with a breakdown of its membership. AICPA has 231,300 members of whom 118,000 are in public accounting and 113,300 are engaged in other activities. AICPA also classified the CPAs in public accounting by the size of the practice as follows:

Individuals	28,200
Firms (2-9 members)	39,800
Firms (10 or more members excluding the 25 largest)	17,800
Firms (25 largest)	32,200
Total	118,000

AICPA also classified its members involved in other activities as follows:

Industry	89,800
Education	6,200
Government	7,600
Retired & misc.	9,700
Total	113,300
Grand Total	231,300

My membership in the American Institute of Certified Public Accountants is dated July 31, 1952. □

CHAPTER X

Auditing Rural-Electric Borrowers

This chapter covers thirty-two years of my life as a Certified Public Accountant in public accounting practice for eleven years as a private practitioner and twenty-one years as a general partner of Schmidt & Company, CPAs. The term "Borrowers" is used for this section because we audited Public Power Districts in the State of Nebraska that qualified for REA loans. In addition to a recital of my trials and tribulations with a few bureaucrats in REA, this chapter is a true story of my experiences as founding partner of a successful CPA partnership.

THE Farmland (CCA) monthly publication *Teamates* dated September 1951 noted this on the front page:

> **WALT SCHMIDT IS CPA**
> Walter G. Schmidt was notified July 20, that he is now a qualified Certified Public Accountant. Walt took the spring exams given in Missouri in May.
>
> He joined CCA as an auditor-analyst in 1942, coming here from LaCrosse, Kansas where he was manager of the Farmers Co-op. He has served as staff analyst in the manufacturing division and is now staff analyst to the controller.

The September 1952 issue of *Teamates* had this article on the front page:

> **OPENS OWN BUSINESS**
> Walt Schmidt, one of CCA's first auditor-analysts, has opened a Certified Public Accounting office at 218 West 73rd Terrace in Kansas City and will be in business for himself after September 1.

As you can see by the dates of the above publications, a year elapsed before I decided to open my public accounting office. At the time W.L. Bradley & Company CPAs of New York was conducting the annual audit for Farmland (CCA) and I investigated the possibility of joining the firm as manager of a branch office in Kansas City. In a letter dated February 18, 1952 Mr. Bradley stated, "I should be very glad to discuss the possibility of your entering public accounting with us at the first opportunity. Of course, if you were to leave CCA to come with us we would have to know that CCA, as our client, was entirely satisfied to have you make the change." Nothing developed from my contacts with Bradley. After I submitted my resignation, CCA did make me an offer of employment at increased pay but by then the die had been cast.

My office was opened at the above location which was also our home. We sold our house as 2901 Forest, Independence, Missouri, September 21, 1950 for $10,500.00 and purchased the property at 218 West 73rd Terrace in Kansas City, Missouri at a cost of $11,000.00. This was about the time my position at CCA was reclassified from the Manufacturing Division under Mr. Zurbuchen to the Control Division under the new controller Mr. Mills. I have fond memories of that home in Independence. The unfinished attic provided space for my den while I was preparing for the CPA examination. That was where I completed the International Accounting Society CPA coaching course studying evenings upon my return from the office and on week-ends. By 1950 our oldest daughter, Paula Kay was attending St. Mary's parochial school. The school bus picked her up at the curb an hour or more before she reached school. She had to suffer the cold when the bus was late. After school she was subjected to the long ride home in a bus not well maintained. Our second daughter, Anita Louise was six years old and was slated to attend the same school. Mother was not very happy with the transportation arrangement. Our third daughter, Nancy Elizabeth, was three years old. It also became obvious that we needed a larger house for our growing family.

In searching for a house Louise and I were looking for one with these qualifications: It needed at least three bedrooms with one of the rooms large. We wanted the house close to public transportation, shopping center and parochial schools. This requirement was based on the assumption that I would travel in my practice and we had but one car. The house needed a full basement for storage and laundry. A modern kitchen or at least one large enough so we could modernize it. The white bungalow at 218 West 73rd Terrace met all requirements and we did remodel the kitchen. The children could walk to school, church, to the bus stop and to the Waldo Shopping Center. The lot was small so lawn-mowing took very little time. St. Elizabeth Church and its fine parochial school were about three short blocks from our house. The street car line was one-half block from our house.

Before one leaves a good job with an annual salary plus fringe benefits, for a profession with no guarantees, one meditates long and hard. I had always said to myself, "If I acquire the CPA certificate, I will try public accounting." The time had arrived to put up or shut up. My plans were to contact members of CCA for any bookkeeping help they might need, to mail announcements to many of CCA's Kansas City employees for possible tax work and to announce a bookkeeping service. One of the replies I received from a former employee working for a management consulting firm in Chicago said, "I hope you have things sufficiently planned so the 'starvation' period will be at a minimum." That did little to build my confidence. The office location in my home was a temporary measure until a suitable office could be located.

In walking to the Waldo Shopping Center, I frequently passed the Crum Realty Company office about one-half block from my home. George Crum, a lawyer by education, told me he came to Kansas City during the depression from a town near Springfield, Missouri. I asked George, "How did you get involved in the real estate business?" He went on to explain that he had worked for a bank as a cashier at seventy-five dollars per month, and being frugal, he was able to save some money. His experience in the bank encouraged him to loan money to worthy borrowers of the bank. When the bank president learned about his banking business, George was discharged. He came to Kansas City about the time of the Roosevelt Bank Moratorium when the banks were closed for a time only to reopen within a few days. Thousands of banks remained closed and were liquidated. George Crum told me that sheriff sales were published for these banks and footlockers of notes and mortgages were brought to the steps of the bank and sold at public auction. Sometimes George and perhaps a half dozen others were the only bidders on the documents. At times, the assets sold for less than ten cents on the dollar of face value. George purchased some of this paper and proceeded to collect payment in the form of cash or property such as chickens, grain or livestock. Profits from these bank liquidations provided him capital that he invested in real estate loans acquired from building and loan associations. This lead to his very successful real estate brokerage firm.

Mr. Crum at one time had a rather large staff of real estate salesmen but when I made his acquaintance he had only two. He learned of my search for an office location and one day offered to share his office with me. The office space amounted to about 2,500 square feet. He said, "All I need is space for myself and the two salesmen. I can have a cabinet-making shop come out and build a nice counter for you with hidden shelving. There would be sufficient room for your secretary, the typing and printing and her supplies. You can occupy the private office and have your private telephone system." By this time George had a substantial investment in real estate in addition to his revenue from the brokerage business and he wanted me to keep his books. We agreed on the terms offered and I rented my first office at 7333 Broadway, Kansas City, Missouri.

My records indicate that I had income from bookkeeping and taxes before officially opening the office in my home. The plans were to open an office at the end of CCA's fiscal year August 31, 1952, so there was no need for secrecy so long as my practice did not interfere with my responsibilities at CCA. The phone rang one day and the party on the other end of the line explained he was involved in housing construction and needed bookkeeping assistance. He said he didn't have time to keep the books and needed help. I told him, "You came to the right place. I have just opened my office and have time on my hands." He gathered his records together and brought them in a box to my office. After piecing the information together and gathering missing items by telephone, I was able to compile a financial statement. It was very evident that his liabilities exceeded the assets of the business. I mailed the statement to him with my invoice in the amount of $200.00 for services rendered. Payment was not forthcoming but one day I received a notice from the courthouse stating that my client filed for bankruptcy. Months later another notice arrived informing me that his assets had been liquidated and a dividend of 75 cents was available at the courthouse.

Unfortunately, the cash receipts for the last four months of 1952 fell short of providing a decent living. By withdrawing contributions made to CCA's credit union and its retirement plan plus loans on my life insurance and small loan on my car, we made it through Christmas. Something else occurred during this time that made all the difference. A good friend of mine, Dale Craven a public accountant, called to relay what he heard at a cooperative accountants' meeting he had attended. The message was that the Rural Electrification Administration had been providing government auditors for its borrowers but was now requiring the borrowers to obtain a CPA or CPA firm to conduct the annual audit of their financial statements.

He said, "Walt I am not qualified to conduct these borrower audits but since you have your certificate, you are qualified." I thanked him for the information.

Since I mailed my announcements to members of CCA and had conducted audits for many of the cooperatives while an auditor, the first contact regarding the audit requirement came from Joe Jeness, statewide manager of the Kansas Electric Cooperatives in Topeka, Kansas. Joe invited me to attend a meeting at the Sunflower Hotel at Abilene, Kansas, where managers and directors of eight Kansas Borrowers wanted to discuss the subject of obtaining a CPA audit. The meeting took place in the fall of 1952. Strangely one of the questions came from an employee of the government as he asked, "In your opinion, why did REA limit the annual audit of the borrower to CPAs?" It was well known that most of the REA field auditors were not CPAs. I told him, "REA probably considered the fact that its borrowers were located in almost every state, the CPAs were licensed to practice in all the states and had promulgated accounting, auditing and reporting standards recognized by SEC and other agencies." Eight cooperatives from these towns in Kansas were represented at the meeting; Axtell, Clay Center, Solomon, Troy, Topeka, Emporia, Wamego and Cheney. The question of audit fees came up for discussion. Some of them said they had served on other boards of directors and were concerned. One of them said, "It just seems like once the CPA gets his foot in the door he begins to turn the crank. The fees accelerate with each passing year." I told them, "My fees will remain the same from year to year unless I notify you otherwise. My policy will be, quality work at a reasonable fee, not necessarily the lowest nor the highest." I told them that regardless of the CPA selected for the audit, since REA required an annual audit of any twelve-month period selected, they would obtain the best audit at the most reasonable fee if they staggered their audit dates during the calendar year. About two weeks later the manager of the Axtell cooperative, R.J. King called and asked, "When can you do my audit?" I asked him, "When do you want it?" After discussing the audit date with the local accountant and determining that a good physical inventory could be taken by November 30, that became the audit date for Axtell and remained the audit date to this day. Pappy, as he was called, was not only a good client but one of my best friends.

Nonprofit cooperatives such as oil, grain or creameries had been my area of expertise but this audit was the first opportunity to audit a nonprofit electric distribution cooperative, a regulated utility in the State of Kansas.

While I was employed at Farmland (CCA) Mr. Cowden, president, returned from a visit to cooperatives in Copenhagen, Denmark, and made this remark. "The rural people of Denmark and Sweden have more electrically lighted hog pens than we have electrically lighted farm homes."

The Center for Research of the University of Kansas has this to say in its book entitled *Kansas Electric Cooperatives—Twenty Years With The REA:*

> *It is probably impossible to determine the exact date of the original central station extension of rural electric service in the United States. However, the approximate date of April 1906 can be used without much fear of being grossly in error. This date marked the initiation of service to rural customers along the right of way of the Aurora, Elgin & Chicago Railway. The*

existence of the power lines, which electrified the railroad running through the rural area, made the extension of service to nearby customers both simple and inexpensive.

In 1935 Congress passed the Emergency Relief Act, which allowed the President to create the Rural Electrification Administration. Executive Order 7037 established the office, set up the organization and defined its purposes. Ostensibly, the purpose was to create facilities for rural electrification, without which the rural areas were being deprived of a modern standard of living. An underlying reason, in addition to this philosophical purpose, was to create a demand for the capital-goods industries in the doldrums of the depression. In turn, more jobs would be created for the large number of workers at the time on the relief rolls and the productivity of the farmers was expected to increase

The Rural Electrification Act was passed by Congress and signed by the President on May 20, 1936. As the bill was passed the REA could make loans to private companies but had to give preference to the cooperatives and public bodies; the interest rate was set at the cost of money to the government, to be determined annually. Reorganization Plan Number II placed the

> *REA under the general direction and supervision of the Secretary of Agriculture on July 1, 1939.*
>
> *In 1944, partially as a result of the low interest-rate policy of the Federal Government and in an attempt to accelerate the rural electrification program the Pace Amendment was passed; this lowered REA interest rates to a fixed two percent and at the same time extended the amortization period of the loans from 25 to 35 years.*

Just as the stock market crash of 1929 resulted in the creation of the Securities and Exchange Commission and fostered the enormous growth of public accounting, the depression of the early thirties resulted in creation of REA and opened a new CPA audit field for 929 electric distribution borrowers, for about the same number of rural telephone borrowers and about sixty or seventy generation and transmission cooperatives. While not a borrower of REA, most if not all statewide associations also engage a CPA firm for their annual audit.

The replacement of government auditors by certified public accountants for REA borrowers took place over a few years. Borrowers in the states of Illinois, Indiana and Iowa were required to engage CPAs before the borrowers in Kansas were asked to do so. The first eight electric cooperatives that I met with in Abilene, Kansas, received their notice in 1952. During 1953 all electric distribution borrowers were required to provide REA with an annual CPA audit report. Working with the American Institute of CPAs, REA published a MINIMUM AUDIT STANDARD BULLETIN as a guide for the CPAs. In addition to meeting the audit and reporting standards of AICPA the audit and the report must also meet the requirements of REA. There were times in my experience when the advice from REA conflicted with the standards adopted by AICPA.

It was my intent to develop a practice that included a broad line of services such as bookkeeping, income tax, management advice, as well as, auditing. Audits were accepted from cooperative creameries, public school districts, REA telephone borrowers and REA electric borrowers, in both distribution and generation. When REA notified the borrowers in Missouri, Kansas, Nebraska, Colorado and Oklahoma to find a CPA for their next audit, my telephone began ringing. REA borrowers have regular national, state and district meetings so news travels fast. In 1953 I conducted audits for seven of the eight REA borrowers in Kansas and from that point forward, the auditing of REA borrowers became a full-time job for me.

My first audit at Axtell, Kansas, and my audit report failed to meet the quality I desired. The REA minimum audit guide had not arrived but the borrower did have a copy of REA's last audit report and the chart of accounts. Most of my office equipment was used and there were no preprinted audit procedures, preprinted internal control questionaires and no preprinted pro forma financial statements. With the influx of REA audits it became very clear that I needed to find a competent audit staff, a full-time secretary and better typing and duplicating equipment. As promised, our landlord had installed an attractive curved counter, I had a private office and we had a nice area for the secretary. An experienced auditor by the name of Bill Hamlett, working for Sinderson & Sinderson CPAs of Kansas City, answered my newspaper ad and became my first staff accountant in 1953. My first full-time secretary by the name of Marian Bayless came to work in 1954. Some desks were purchased from Mr. Crum, a Model 80

Multilith printing machine and a new IBM electric typewriter were purchased for the office. Portable typewriters and calculators were purchased for the audit staff.

As my name became known, invitations began arriving from officers of statewide REA accountants to appear on their annual meeting programs. REA auditors and field personnel were always invited to attend. While the annual audits of REA borrowers has been the responsibility of CPAs, REA continued to retain some of its auditors to the present time. We have been told by clients, "They spend more time auditing the CPA than they do auditing us." One such auditor was at an accountants meeting in Scottsbluff, Nebraska, and told me that he was leaving REA to work for Peat, Marwick, Mitchell CPAs. He explained that REA wanted him to switch from auditing electric borrowers to auditing telephone borrowers, an assignment he did not accept. I asked him, "Why would you seek employment with such a large international auditing firm instead of a smaller regional firm?" He told me, "Well, they made what I considered a good offer and I accepted." Whereupon I told him, "If PMM doesn't treat you right, give me a call." That call came years later and he worked for me for three years. To my knowledge Peat, Marwick, Mitchell was the only large auditing firm actively seeking the REA borrower audits. A partner in their Omaha, Nebraska office was hiring some REA auditors holding CPA certificates. A CPA by the name of Baker entered private practice in Iowa after leaving REA and another man by the name of Tarr entered private practice in Colorado after leaving REA. Both were subsequently hired by PMM. They were tough competition while in private practice. At one of the REA Accountants meetings in Grand Island, Nebraska, I asked Mr. Tarr, "Why did you leave private practice? You were well liked and had a strong following among the REA borrowers." He told me, "Walt, the travel and work became a real burden. Both my wife and I worked night and day and were unable to find a suitable office or to find competent accountants for my staff. PMM has a nice office in Denver, a good staff and coupled with the offer they made, I accepted."

For the first five years or so I conducted as many of the new audit engagements as possible and Bill conducted the second and third audit. In breaks between audit engagements, Bill worked on statement work, taxes, etc. Our secretary had been told to send tax clients to neighboring CPAs in Waldo when Bill and I were out of town on audits.

It has always appeared to me that if at all possible, a person should be compensated in proportion to his accomplishments or production. In many cases there is no good measurement of a person's contribution to his employer and a salary is about the only alternative. My auditing practice became regional, very specialized and the audit staff traveled extensively. I paid my auditors on a commission basis or a percent of revenue produced. Their travel reimbursement was charged against earnings. It served me well and I never received a complaint from my auditors. The office secretary and the typists were paid on an hourly basis. While my office was at 7333 Broadway, my wife and my daughters helped my secretary as the work load increased. None of them enjoyed operating the Multilith machine.

The second five years brought more REA audit engagements and I employed another accountant to assist with the audits and to handle the local work consisting of bookkeeping, taxes, etc. When I began auditing REA borrowers, REA encouraged employees of the local cooperative to take a bookkeeping course compiled by REA. Many of the employees enrolled in the course and resulted in a better set of records for us to examine. In one of REA's austerity binges it was decided to drop the course. I contacted REA and obtained permission to update the course and publish it under our name. Permission was granted and I revised the course and offered it to our clients for $200. After we had sold a number of the courses, consisting of two loose-leaf binders, REA decided to reissue the course and offered it to its borrowers for $85. Needless to say, we discontinued selling the service.

From September 1, 1952 to May 31, 1963 I practiced public accounting as an individual maintaining a local practice but more and more becoming specialized in the audit of REA borrowers. The CPA profession has local chapters that meet at regular intervals. At one of those meetings I met Paul Schmidt, no relative, who had acquired partnership in a local firm. He made the comment that two Germans with the same name should belong to the same firm. One day I had a call from Paul and he said "We have been kidding about belonging to the same firm and now I am faced with the loss of my partner. He had a stroke and will not be able to continue. Are you interested in a partnership?" I told him that I had given considerable thought to a partnership, if for no other reason, than to provide for continuity of the practice. We exchanged information regarding our practices and decided a merger might be beneficial to both practices.

It was evident that we would have to find a larger office to accommodate both staffs. We agreed that we would operate by two divisions, he would maintain his audit staff and clients, render billing and maintain his own records. I would do the same for my staff. It was agreed that I would be the partner designated as office manager and would supervise the secretary, who in turn would supervise the typists. It was also my responsibility to arrange for a custodian. Direct expenses in connection with the audit staff, such as travel expense was to be paid by each partner. Indirect expense such as office labor, utilities, telephone, office supplies, microfilming, tax library, etc. was to be paid by me and at the end of each month Paul was to reimburse me for his portion of the office overhead. The formula for the allocation was to be based on a detailed analysis of the previous calendar year's experience approved by both partners.

Our next problem was to find suitable office space. Paul wanted an office near downtown where many of his clients were located. I would rather have remained out of the downtown area. We finally found a building at 2837 Main Street that appeared suitable but the owner was a life insurance company of Dallas, Texas, and did not want to lease the building, they wanted to sell. Neither Paul nor I had much available capital but the real estate agent urged us to make an offer. We finally offered them $52,000.00 for the building with 10% down and five percent interest payable in equal installments over twenty years. It took them about a month before accepting our offer. At Paul's suggestion we formed a new corporation under the name of Schmidt Investment Co., Inc. to purchase the building. Each family owned $2,600.00 in common stock.

The building provided us with about 2,400 square feet on the first floor and the same on the lower level minus space occupied by heating and air conditioning equipment. It had no private offices, so we hired a cabinet maker to build us suitable credenzas, two long counters and two private offices for the partners. The partnership of Schmidt & Company, Certified Public Accounts, was organized June 1, 1963 at 2837 Main, Kansas City, Missouri. As soon as the office was ready for occupancy we moved in. The partnership leased the building from Schmidt Investment Co., Inc. at rent adequate to finance the acquisition and provide for taxes, insurance and maintenance. Rent was prorated between Paul and I on the basis of space occupied.

The REA managers of Kansas held their semiannual meeting in Wichita and invited me to prepare a paper on the desirability and feasibility of converting their consumer billing from the billing machine to data processing on IBM mechanical punch-card equipment. A meeting was arranged and the Director of the Kansas Corporation Commission was invited. My intent was to use the lower level of our building for a data processing center for our clients. After preparing some financial forecasts and procuring the assistance of IBM, I made a presentation to the managers and several of them indicated they were ready to switch their billing to the IBM punch card system. I rented the IBM equipment and planned to have the service available for January 1964 billing for seven cooperatives serving about 25,000 customers. Rural Data Processing opened for business in January 1964 after I and my family worked long hours to meet the deadline.

Rural Data Processing (RDP) was successful from the beginning because it was designed to provide the cooperative with services needed. Those included billing, control of cash, aging of accounts receivable, delinquent notices, member and nonmember business, customer identification by map location and system location, load studies, rate studies and allocation of capital credits at the end of the calendar year. To my knowledge this had never been accomplished automatically. When IBM learned of my interest in providing this service for our clients, two young gentlemen in dark suits, white shirts and black ties came with notebook in hand stating, "Mr. Schmidt this is IBM's complete system for the rural electric cooperative." I told them, "I know all about your system but for Rural Data Processing you might as well put your book aside. I will tell you what our clients need. I will suggest what should be punched in each of the eighty columns of the IBM card and you gentlemen will have to tell me if you can wire these machines to give us the reports we need." RDP succeeded because IBM and I worked together to design the system. It was a far-reaching system that took the farmer at the end of the line, the cooperative in the rural town and RDP in Kansas City working together to get the job done. RDP had to furnish the farmer with prepunched IBM cards and instructions to use them and supply the local cooperative with a master file of customer cards for their use. At regular intervals the local cooperative would accumulate the cards received from the farmers, balance the cash receipts, attach an adding machine type to the cards and mail them to RDP. Using IBM key punches, verifyers, collators, sorting machines, calculating punches and tabulators, RDP would process the information received and mail printed reports back to the cooperative. RDP charged a monthly fee for its service on the basis of the number of customers served during the month.

As the success with my Bookkeeping Course brought competition from REA when the course started to sell, so my adventure into the data processing field invited competition. The National Rural Electric Cooperative Association (NRECA) of Washington, D.C. took note of RDP and the glowing reports it evoked and encouraged the cooperatives to organize a number of these data processing centers on a much larger scale. A memo was mailed to cooperative managers endorsing such a move. A manager of one of our clients who was not using RDP sent me a copy of his letter to NRECA regarding the memo:

> *This letter is with regard to your memo to Rural Electric Managers and*

Statewide Managers, dated December 30, 1966, having to do with data processing.

Both NRECA and REA for some reason continually ignore a data processing operation that I feel is the best in the country and will continue to be the best in the country, bar none.

I refer to the operations that Walter G. Schmidt, CPA partners of Schmidt & Company, Certified Public Accountants, 2837 Main Street, Kansas City, Missouri 64108, started some years ago and has a successful operation. Mr. Schmidt knows the rural electric business and knows the answers that must be obtained through data processing and knows how to go about getting the answers.

Everybody seems to give a big play to the operation up in the Dakotas but in my opinion they have not proved anything yet, and I do not think they know the rural electric business well enough that they can ever have a successful operation. I realize this is a strong statement, but it 'burns' me every time people try to push an organization like that, and ignore an organization that knows what they are doing.

We have not converted to data processing in any manner. Perhaps I am of the old school, but I feel everything that can be

> *done on premise should be done there, for control purposes. If we go to data processing outside our own organization, we will certainly go to a firm that knows what they are doing.*

When RDP went into operation it also alerted all manufacturers of punch card equipment or computer hardware to new sales prospects in the rural areas of the country. A number of local managers concluded they could buy or lease equipment to perform automatic data processing on a local basis. One of our audit clients had such a system and as their CPA it was up to me to call attention to the system's shortcoming. A year later I learned that the local manager accepted my comments, not as coming from a CPA, but from RDP, a competitor. Schmidt & Company lost the annual audit.

The heavy auditing schedule for the last quarter of 1965 left little room for what was about to occur. I became ill one evening at home and my family took me to the emergency room at St. Mary's Hospital. The doctor on duty took very little time as he held his hand to my side and said, "You have a hot one here" referring to my gall bladder. A surgeon was called and surgery was scheduled for the following morning. Later that night the surgeon came to my room and warned that the surgery might be performed in two steps. Depending upon the degree of infection, the first operation would remove the stones, but if deemed necessary, I would be sent home with some medicine to return about three weeks later for removal of the bladder. This was what took place in my case and I was not able to travel until December of that year. I had my secretary send a note to the audit clients on my schedule to notify them the audits would have to be rescheduled. Some of them called and said, "Don't worry about the audit. Do it when you are able. REA didn't get around to audit us for years."

The auditing practice was growing dramatically as the years passed and RDP was acquiring new customers for its services. RDP employed about ten people and rendered customer billing for over 40,000 rural customers. In addition to my medical problems in 1965 personnel problems also arose. By this time my staff included three auditors and the four of us were conducting nearly one hundred audits annually for REA electric and telephone borrowers. In the summer of 1965, Vertis Ambrose, one of my auditors came to me and said, "Walt, we just learned that one of our boys has developed serious coronary problems and we will have to watch his activity closely. This will add to the responsibilities at home and I feel the extensive travel required by Schmidt & Company deprives my family of my assistance. I have found other work allowing me to be at home." Vertis was well liked by the clients and his work was excellent.

While I was in and out of the hospital in the fall of 1965, another of my staff, Roscoe Quint, CPA, contacted me and demanded that Schmidt & Company accept him as a partner. Roscoe passed his CPA examination while he was working for Schmidt & Company. He also took care of our local practice consisting of bookkeeping, taxes, etc. His ultimatum was either acceptance as a partner or he would resign and start his own practice. Roscoe asked, "What are you going to do about it?" My first reaction was one of disbelief. After regaining my composure, I faced Roscoe with my answer, "Roscoe, you were the one I was grooming as my replacement but you couldn't wait. Vertis's resignation was understandable but your demand at a time like this is blackmail and despicable. I know exactly what I am about to do. I will write your final pay check." Roscoe Quint opened his own practice and took a few of our local clients with him. We had a noncompetition clause in our written employment contract and I threatened to file suit for a permanent injunction to keep him from serving former clients of Schmidt & Company. We settled out of court when Roscoe signed a note payable to Schmidt & Company for the annual revenue we lost when the clients retained him. The installment note was paid in full. New accountants were hired and Schmidt & Company retained all of its REA audits.

When Schmidt & Company was formed in 1963, the office at 2837 Main Street provided space for the combined auditing practices of both partners and almost an entire lower level for RDP. With both of our practices and the services rendered by RDP increasing in size, we were finding ourselves cramped for office space. My partner Paul was performing bookkeeping services on an old Frieden computer and needed more modern computer hardware. Furthermore, having the computer and the printing equipment on the same floor created a noise problem.

The resignation of one of my CPA employees and the subsequent loss of some local bookkeeping and tax clients made me question the value of a nonaudit practice commingled with an essentially specialized audit practice. The reluctance of audit clients to accept suggestions regarding their computer application, inasmuch as Schmidt & Company had its own data processing center, was also a concern. It appeared that our specialized audit practice with REA borrowers took so much of my time and effort that serving local bookkeeping and tax clients was in the best interest of the auditor serving those clients and not in the best interest of Schmidt & Company. In the case of RDP, a successful automatic data processing center, owned and operated by a CPA firm made it difficult for Schmidt & Company to evaluate client installations.

While my revenue from auditing was by far the major source of income, this was not true for Paul's half of the partnership. His largest revenue producers were compilation work or bookkeeping and taxes. I contacted Paul to see if he would be interested in purchasing my local practice since it would merge beautifully with his major practice. We reached an agreement and I turned about one hundred local clients over to him and he paid me over a period of three years as services were rendered for the clients. This left my half of Schmidt & Company with an audit practice for REA borrowers but still rendering data processing services.

It was rather obvious that data processing on punch card equipment was being replaced by electronic computer equipment and considering the possible conflict of interest question with our clients, I approached the cooperatives we were serving for a possible sale of RDP to them. I reasoned if this could be accomplished diplomatically, it might solve our conflict of interest problem with our audit clients and our office space problem at the same time. Another event that perturbed me and created tension between Schmidt & Company and one of its audit clients, was when a manager using RDP services hired RDP's manager as his office manager at a substantial raise in salary.

RDP's customers were receptive to the idea and called a meeting to form a nonprofit corporation. An attorney was engaged to draft Articles of Incorporation and by-laws for the Rural Data Processing Association, Inc. Since RDP was serving cooperatives in several states the matter of location had to be resolved. Most of the customers resided in Kansas so Topeka, Kansas was selected as the headquarters for the new service corporation. Charles W. Ellis, manager of the cooperative at Clay Center, Kansas was elected president. Capital had to be raised by the new corporation and a purchase and sales agreement had to be negotiated, satisfactory to me and the new corporation. My attorney drafted a purchase and sales agreement and a bill of sales agreement in line with agreements reached during negotiations. On April 25, 1973 the president of RDPA and I signed the agreement. The closing of the transaction was to take place January 2, 1974 at 2837 Main, Kansas City, Missouri. My records indicate RDP had a gross revenue of $113,203.13 for the calendar year 1973. While most of the equipment was leased from IBM, RDP owned equipment costing $15,950.57 and an inventory of supplies costing $4,616.21 at closing date. Most of the equipment was ten years old. I agreed there would be no break in services rendered for the remainder of 1973 and the new corporation would have until January 31, 1974 to remove its assets from the lower level of Schmidt Investment Co.'s building at 2837 Main. The purchase price of $50,000.00 was delivered to me on January 2, 1974 in the form of a cashier's check and RDPA removed its equipment and supplies during January 1974. In addition to retaining all customers of RDP as audit clients, Schmidt & Company fell heir to a new audit client for the fiscal year ending September 30 by the name of Rural Data Processing Association.

Perhaps our CPA profession will eventually learn the lesson I learned about twenty years ago. Not only are the large CPA firms involved in audits of public companies and suffering from adverse court rulings in malpractice cases but the public is questioning the independence of the CPAs. When CPA firms develop software programs, assist in overall corporate planning, market research, tax strategies, executive recruiting, compensation packages and identifying prime acquisition candidates, are the CPA firms independent or are they working for the client? It has been said that fifteen percent of the Big Eight revenue comes from their Management

Accounting Service (MAS) division. It includes General Management, Financial management, Operations management, Marketing, System management, Public utility services and Human services.

Since my first audit of an REA borrower in 1952, I was responsible for the annual audit of over a hundred electric and telephone borrowers and I must have signed over fifty thousand audit reports before I completed my retirement from Schmidt & Company at the end of 1984. The clients were scattered over eight midwestern states from Wyoming to Louisiana. My career as auditor for REA borrowers has been interesting and rewarding. It provided me the opportunity to become personally acquainted with thousands of individuals serving on boards of directors or as employees of the clients. I can truthfully say, serving these fine people over thirty-two years was most enjoyable and I regard them as personal friends. There were very few disappointments in our relationship with the client.

One case comes to mind that was a real (Believe-It-Or-Not). The IRS made an examination of one of our clients and determined the client failed to meet the requirements for exemption from income tax due to nonmember business activity over a period of years. A tax assessment was made by IRS in the amount of about two million dollars plus interest of about one million dollars. The cooperative continued to file IRS Form 990 claiming exemption and IRS postponed legal action by requesting extensions of the statute of limitation. I did the audit personally and the client kept me informed regarding the case. The client finally hired a local law firm and an outside tax attorney recognized as an expert in the field. During one of the annual audits I discovered the Board of Directors was seriously considering accepting the advice of its lawyers to pay IRS about $800,000.00. During my oral report I told the board, "If it was my money, I would see IRS in court before I paid the $800,000.00. Litigation costs money but that is a lot of money in my judgment. I do not believe you owe IRS that much." The board advised the attorneys not to settle but to proceed to the tax court.

The attorneys notified IRS and proceeded with their case by taking depositions from people involved and maintaining contact with IRS. I prepared many schedules of figures to support our case and supplied them to the attorneys. The date was set for our appearance in court. While I was in Estes Park, Colorado, on my way to a client in Colorado, a phone call from one of the attorneys reached me. The attorney said, "Walt, IRS has agreed the cooperative meets the requirements for exemption for all but one year and is willing to settle for $100,000.00." Whereupon I told the attorney, "Now settle, we can't afford to go through court for that amount." I had audited this client for about twenty years at a very reasonable audit fee and after saving the cooperative $700,000.00, I was certain the cooperative would be a contented client for a long time to come. After the settlement I received a letter from the tax attorney in which he stated, "Without your determination and persistence we could not have reached this settlement with IRS." What was my client's reaction? The client engaged another CPA firm for the annual audit the following year.

Another distasteful episode involved the integrity of a manager employed by two electric cooperative clients. The manager terminated his employment with one cooperative and accepted the manager's job with another at increased compensation. Subsequent to the transfer, our audit of the financial statements of the first employer discovered a significant embezzlement perpetrated by the manager while so employed, a crime to which he confessed. In our audit report to his new employer, we alerted REA and the Board of Directors as to our findings. Instead of both parties asking the manager to submit his resignation, they approved the engagement of another auditing firm.

During the seventies, some of our clients were being contacted by rural banks and encouraged to purchase U.S. government securities from the bank at a price and the bank by executing a repurchase agreement, was to repurchase the securities later at a higher price. One of our clients purchased securities in this manner and the securities were held in a custodial account by a third party. In that case the financial statements reflected the investment in U.S. treasury securities. In all other cases, where the bank retained possession of the securities, our policy was to treat the transaction as an unsecured loan to the local bank. Failure to treat the investment in this manner resulted in a qualified audit report. One of our clients changed audit firms due to our position on repurchase agreements. All of this took place before the news of the bribery and fraud scandal broke regarding the repurchase agreement transactions by the E.S.M. Government Securities, Inc. of Fort Lauderdale, Florida. In that case, investors including scores of savings institutions in Ohio, municipalities, schools, etc. lost hundreds of millions of dollars in their investments in repurchase agreements. One of the "Big Eight" auditing firms and its insurance companies are facing staggering malpractice litigation claims.

Conducting audits for cooperatives being financed by an agency of the federal government was not new to me. While I was auditing for Farmland (CCA), many of the local cooperatives were borrowing money from the Banks for Cooperatives under the supervision of the Farm Credit Administration.

Copies of our audit report, although not CPA and not independent, were mailed to the Bank for Cooperatives. In fact, a suggestion from a vice-president of the Wichita Bank for Cooperatives is part of Schmidt & Company's reports. In a presentation at an accountants' meeting he said, "When you auditors refer to an exhibit in your report, why don't you tell me what page it is on?" There was never any doubt about the relationship between the parties involved. It was a banker-borrower relationship. I assumed this would be the relationship between the Rural Electrification Administration and the borrower. In the following pages you will see how wrong my assumption was.

The Rural Electrification Administration, acting for the federal government, loans money to the local rural electric and telephone cooperative and holds a first mortgage on the assets of the cooperative as collateral. There is a lengthy mortgage agreement between the parties specifying actions by the cooperative requiring REA approval. Most of REA's legal power over the cooperative comes into play only when the cooperative fails to meet its debt service payments to REA. One of the obligations of the Board of Directors of the local cooperative is to engage a CPA for an annual examination of its financial statements. While this is subject to approval by REA, only once in my thirty-two years of auditing did REA oppose a local board resolution engaging me or Schmidt & Company to conduct the audit. A new resolution is required for subsequent years only if the cooperative changes audit firms.

REA has been organized and reorganized many time since 1936 but generally speaking it administers its program starting with the Administrator in Washington, through his deputy and then through a number of area offices that implement the financing, engineering and at times the auditing of the borrowers segregated into areas of the country. REA is under the Secretary of Agriculture who is appointed by the President. The REA Administrator serves under the Secretary of Agriculture and is also appointed for his term of office. When political parties change in Washington, faces change at REA. Periodic audits of borrowers until 1947 were conducted by government employees under their chief in Washington. Begining with the Truman Administration, selected borrowers were asked to engage a CPA to perform annual audits formerly made by REA through its field examiners. In a letter to all electric borrowers dated September 2, 1953, R.G. Zook signing for the Administrator made this announcement:

> *In the fiscal year ended June 30, 1953, 634 borrowers were responsible for their own audits. Based on experience, it now appears appropriate to ask all electrification borrowers to undertake the employment of independent certified public accountants to audit their accounts during the fiscal year 1954 and each succeeding year. The audit prepared by the certified public accountant will replace the annual audit formerly made by REA through its field examiners.*

I opened my office September 1, 1952 and in my file is the following letter from REA dated November 17, 1952:

> *This is in reply to your letter of October 21, 1952, in which you request copies of various publications.*
>
> *We do not have the publications listed for general distribution but suggest that ac-*

> *countants arrange to use copies which have been furnished the REA borrowers by which they may be engaged.*
>
> *We do have a very limited supply of the Uniform System of Accounts and the Manual on Work Order Procedure, and are following the practice of providing, upon request, one copy of each to firms which have been selected by one or more REA borrowers to perform their annual audits. Upon your advice that you have been selected by a borrower, we shall be pleased to send you a copy.*
>
> *Division of Controller*

On November 22 I notified REA that the REA borrower at Axtell, Kansas had engaged me to conduct their audit and requested they send me the available bulletins. REA's reply dated December 2, 1952 was as follows:

> *In response to your letter of November 22, we are enclosing the following:*
>
> *Uniform System of Accounts Prescribed for Electric Cooperatives Financed by the Rural Electrification Administration.*
>
> *Manual of Work Order Procedure and Related Instructions on Force Account Construction by REA Borrowers.*
>
> *Very truly yours*
> *Leslie Surginer, Controller*
> *Division of Controller*

From my first contact with REA, the government has retained career government employees who prescribe audit standards for the CPA, review the audit reports submitted by the CPA and reserve the right to examine his work papers.

A memo dated August 26, 1953 from the administrator to all REA Employees contained the following:

> *In the interest of efficiency and economy certain of our activities are being discontinued, curtailed, or transferred to borrowers. As a result of these actions a small reduction in force is being made at the present time.*
>
> *It is expected that 48 employees will be affected by this reduction. Of this number, 37 will be separated and 11 will be reassigned to lower grade positions.*
>
> *Employees affected will have received their notices by the time this memorandum is distributed.*
>
> *I am deeply concerned about employees who are affected by the actions resulting from reduction in force. I have asked our Personnel Division to do whatever is possible to assist them with their individual problems and in securing other employment.*
>
> *I wish to emphasize that those to be affected have been determined strictly in ac-*

> *cordance with the regulations and in no case is the action intended to reflect on the ability of any individual.*
>
> *Ancher Nelson*

At one of the gatherings of the REA borrowers, I heard a representative of REA make the statement, "At one time REA had nine hundred employees. Now our staff has been reduced to seven hundred and by golly we intend to get them back." I mention this because it may help to explain the frustration I experienced in my efforts to please a few key people in the audit review section of REA. Some of the REA representatives in the field, both loan people and accounting people, were actually praising my work in their contacts with the borrower. Unfortunately, just the opposite position was taken by a few accountants in the Washington office. We may never know what prompted the discrediting of the CPAs and particularly one CPA by the name of Walter G. Schmidt. I have my own theories. While meeting with REA representatives at area accountants' meetings I received the impression that REA accountants resented the CPAs rendering a service they had been rendering for fifteen or sixteen years. I had breakfast with Leslie Surginer, controller of REA, at one of the Missouri REA Accountants' annual meetings in St. Louis. Mr. Surginer was a CPA and told me he aspired to entering the public accounting profession at one time, had his stationery and calling cards printed and opened his office. The practice did not develop as he had hoped so he sought employment elsewhere. John Scott of REA attended many of the accountants' meetings and early in my contacts with him he was Assistant Chief Electric Operations and Loan Division. In discussing the auditing program at one of the meetings he thrust this question at me, "Walt, why don't you do what we tell you? You can't afford to fight the federal government." Why is it that so many bureacrats in Washington would like for us to believe they are the federal government? My answer to John's question was, "For one thing, your minimum standard bulletin requires me to follow the standards promulgated by the American Institute of Certified Public Accountants and for another thing, I am not working for you." I soon learned that some REA employees do not take kindly to opinions contrary to their own. On another occasion I was invited to the Missouri REA Accountant's meeting in Jefferson City, Missouri and the president of the accountants asked an REA field accountant, "As office managers, we have no authority regarding the co-op investments but we are asked about the security of investments in FSLIC insured accounts, accounts insured by FDIC, FHA notes or in treasury securities, I would like for you to clarify the difference in these investments?" His response was to the effect that all were approved by REA and guaranteed by the government. I had just entered the room before the question was asked so the president of the club turned to me and said, "A CPA just entered the room, let us get his opinion regarding these investments." I told them that of the four only the treasury securities were direct obligations of the United States and went on to explain the origin of the Federal Savings and Loan Insurance Corporation (FSLIC) and the Federal Deposit Insurance Corporation (FDIC). I also discussed the nonliquid risk in the purchase of FHA notes assigned to the cooperative. This was the information sought but I am certain it did not endear me to the REA field accountant.

In the conduct of my first audit at Axtell, Kansas, about the only reference material available was the Uniform System of Accounts and a copy of the last audit report prepared by an REA auditor. It was a rather short report but supplied the necessary financial information to the board of directors and included suggestions for management improvements. Part of the problem was my audit was conducted within weeks after I was engaged to perform the audit. According to my files the first Minimum Audit Standards Bulletin was issued in 1952 but a more comprehensive bulletin was issued in 1956. As a sizable number of my audit reports arrived at REA, two copies of the report were required, copies of audit review letters addressed to the President of the local board of directors began arriving on my desk. The letters contained exceptions taken by REA to my report. The letters were signed by the head of the REA area in which the client was located. I contacted REA and voiced my consternation at REA discrediting my report in their letters to the clients. It was my contention, the client had nothing to do with the manner in which the report was prepared and any criticism of the report should be addressed to me and not the client. By having the Area Director sign the letters it also shielded the Audit Review Section that drafted the letters. I have in my a file a letter directed to me dated May 23, 1958 referring to my audit report of a Missouri client. Here are some excerpts from the letter:

> *In our review of the audit reports prepared by you we have noted in a number of instances that Exhibit C is not prepared in accordance with the sample reflected in REA Bulletin 185-1.*
>
> *The items to which we are specifically referring are as follows:*
>
> > *1. Your reports reflect but two margin categories, namely, operating margins and non-operating margins.*
> >
> > *The REA Uniform System of Accounts provides for five (5) margin and as-*

> *sociated accounts under "Other Member and Patron's Equities." Therefore, if the records reflect balances in one or more of these equity accounts, they should be reported separately on Exhibit C.*
>
> *2. Your Exhibit C reflects what appears to be some type of reconciliation or breakdown of margins by segregating amounts to a category entitled "Capital Credits." However, based upon information contained in the report and the figures shown on the exhibit, it is impossible, in most cases to determine how you arrive at this breakdown.*
>
> *From your exhibit, and the audit comments, we are unable to determine two very important facts, which are:*
>
> > *1. The amount, if any, of margins that have been actually assigned, and*
> >
> > *2. The year or years' margins have been transferred to the assignable or assigned account.*

We therefore request that, in the preparation of future audit reports for REA borrowers, Exhibit C be prepared in a manner conforming more closely with the example in REA Bulletin 185-1.

My reply to the above letter contained the following excerpts:

> *Dear Mr. Askegaard:*
>
> *Capital credits, when it appears in my audit report means just what account number 201.1 defines, 'This account shall include the amounts of patronage capital which have been assigned to individual patrons as capital credits.'*
>
> *Mr. Askegaard, when I glance through our correspondence file regarding REA audits, I can't help but feel that both the REA Administration and my office are spending entirely too much time expressing differences of opinion regarding the reporting of members' equities. In the first place neither the REA Administration nor I can do much about it. All we can do is advise and then the borrower does as it pleases. We find almost as many ways of handling the problem as we have clients.*

On June 3, 1958 I wrote to Richard Richter, Head Operations Section:

> *A few weeks ago Bill Lines of REA suggested it would be time well spent for me to visit with you sometime. Last week John Scott called me and invited me to come to Washington to meet people in the REA Administration in the hope that we can elimi-*

> *nate some of the disagreement regarding the audit report. I am writing Mr. Scott and suggesting the second week of July as a desirable time for me to go to Washington. Will your program offer an opportunity for a short meeting during the week?*
>
> *Of one thing you can rest assured. My record over the past twenty-five years proves that I am a friend of farm cooperatives. My record also proves that I endorse wholeheartedly the REA program and am ready to defend it at any time. My only aim is to render a valuable service to both."*

A meeting was arranged for July 11. Upon my return I wrote a letter to Mr. R.G. Zook, Assistant Administrator of REA dated July 29, 1958:

> *I had a nice trip to Washington and I especially enjoyed meeting many of the people in REA whose responsibility it is to administer the REA program. We had a nice discussion regarding the CPA audit reports, as well as, other problems in connection with the borrowers. Out of it should come a better understanding of our mutual problems. Important as audit reports are, we must never let them take preference over a competent examination or adequate disclosure of the circumstances even though the disclosure may not comply in form with REA accounting suggestions or reporting samples.*

Some of the items listed in your letter as "conclusions reached" are actually a confirmation of report practices that we have had in effect for the past several months. Some items represent the same information presented in a different manner for easy reference. Others represent more detailed reconciliation of accounts and more elaborate disclosure of auditing procedures employed. I realize fully that our services are valuable only if they supply both the borrower and the REA Administration with information essential to the program. Any suggestions for the improvement of our audit report are appreciated.

There was no agreement to change any exhibits used in our report except that I agreed to add information either in the exhibits or in the comments to more nearly comply with REA Bulletin 185-1. It was my opinion at the meeting and remains my opinion that material suspense accounts have no place on the balance sheet and that patronage refunds from a generating cooperative must be reflected on the revenue and expense statement of the recipient cooperative to comply with accepted accounting principles. The only time a deferred charge is justified on the balance sheet is if it can reasonably benefit the future. I fail to see anything wrong with our opinion paragraph. The accepted opin-

ion according to the American Institute of CPAs is as follows:

> *In our opinion, the accompanying balance sheet and statements of income and surplus present fairly the financial position of X company at December 31, 19--, and the results of its operations for the year then ended, in conformity with generally accepted accounting principles applied on a basis consistent with that of the previous year.*

This is what we refer to as an unqualified opinion and is the form we use except the word 'income' is replaced by the work 'equities' to reflect the nonprofit character of the corporation.

It seemed like there was no way to stop the torrent of letters from REA attacking and criticizing my audit reports. Most of the letters were directed to the borrower with copies to me while some came to me directly.

My experience during the fifties was under the Eisenhower administration, with Dave Hamil, REA administrator. With the general election of 1960 the Kennedy-Johnson administration came into power and for the next eight years the REA program was under the direction of a new administrator. I wrote the new administrator, Norman Clapp on March 17, 1961 and received a reply in which he stated, "We appreciate your sincere interest in building up the financial stability of our borrowers." Unfortunately, the change in administrators meant a reshuffling of government employees but no improvement in their attitude toward the CPAs or my work. The situation kept deteriorating and my clients were getting concerned. Kansas REA clients that trusted me with their first CPA audit decided to take matters into their own hands. The regional NRECA meeting was being held in Wichita and seven of these clients asked Mr. Clapp for a meeting with them and myself at a time convenient for him during his stay in Wichita. The clients wanted their CPA and their banker on the same team. R.J. King, co-op manager at Axtell, Kansas was known for his candidness and it didn't take long for it to appear. When Mr. Clapp asked him for his comments regarding the disagreements with the audit report he said, "Mr. Clapp when REA was doing our audit we received perhaps one original copy of a typed report containing a few pages and three carbon copies. The original copy was the only legible copy so there were no copies for the directors. Now we receive many printed copies of the CPA report, containing much more information and each member of the board has a legible copy. Mr. Clapp, you're just not in the same league." Mr. Clapp being a good politician listened to their endorsement of the CPA program but made no commitments with respect to correcting the insidious campaign by his accountants in Washington.

Another meeting was called at Grand Island, Nebraska by some of my clients inviting me to come to a meeting at which some of my clients and Ralph Foreman, Deputy REA Administrator would be present. These clients were also disturbed with REA criticism of our report. During the meeting I said to Ralph, "We probably have one of the nicest reports coming into REA and yet our report receives this relentless abuse from your accountants." He replied, "Walt, there is no question about that but I must warn you John Scott can be ruthless if he decides to take after you."

There was really no way for me to tell who prepared the letters but the key people in the audit review process under all administrations were Leslie Surginer and John Scott. Both of them from Texas and I was told had the ear of a very powerful democrat in congress, Sam Rayborn from Texas.

On February 5, 1962 I received an invitation from Bob Kabat, Director of Management Services of NRECA including this message:

> *A number of managers have asked us to present this workshop just for CPAs who work with rural electric systems. This will be a technical presentation and delve in depth into some of the accounting problems involved, as well as, the particular areas you might be called on for advice and assistance. In other words, there will be this one offering of Workshop VII during the year 1962 tailored to your needs. It will be held on Friday, Saturday and Sunday, March 2-4, 1962 at the Chalfonte Hadden Hall Hotel, Atlantic City, New Jersey, just prior to the NRECA annual meeting.*

He mentioned in his invitation that a team of six people would conduct the program. One of them was John Scott, Assistant Administrator, REA and another was Leslie Surginer, Controller, REA. NRECA is an abbreviation of National Rural Electric Cooperative Association.

I accepted his invitation and in my acceptance letter made the following remarks:

> *The meeting will be a little strained because of the attitude of some of the REA auditors towards CPAs. This group has been conducting a vicious propaganda campaign against the work of CPAs for the past several months. I cannot speak for other CPAs, but critical audit review letters directed to my clients are obviously designed to discredit the CPA.*

A few days before my flight to Atlantic City I received word that REA had sent a letter to my clients inviting them to two meetings at which W.G. Schmidt, CPA would be present at the Chalfonte Haddon Hall Hotel. The meeting with the Kansas clients was to take place on Sunday afternoon. I received a notice to meet with REA representatives at five o'clock Saturday evening. I took an Eastern flight to Kennedy airport with connections to Atlantic City with some trepidation. When my flight arrived at Kennedy there was no connecting flight to New Jersey because personnel and equipment at the airport were busy fishing people out of the harbor. A jet liner had plunged into the harbor on its takeoff. Eventually, transportation became available and I made it to Jersey City but my bags did not. I waited for hours before they finally arrived but my flight to Atlantic City left long before then. The last leg of my trip to Atlantic City was by bus occupied mostly by maids returning from work. My reservation was at a motel across the street from Hadden Hall where the NRECA annual meeting had its headquarters. Luckily there was no problem with the motel reservation.

The CPA workshop opened on schedule and about twelve CPAs from various states were present. The one closest to Kansas City was one from Iowa. A CPA from the state of

Washington probably made the longest trip. The program rolled along rather smoothly with the REA personnel expounding their requirements and the CPAs listening, for the most part. I purposely refrained from raising the problems REA and I had. The program moderators spoke in generalities and also avoided the issue. At one point in the program I passed a note to the CPA seated next to me and asked him to pass it on. The note mentioned that REA was calling a meeting of my clients Sunday and Monday but as a prelude they wanted to meet with me Saturday evening at five o'clock. The note gave them my room number and invited them to come to my room after five if interested. After the meeting I converted my room to a mini hospitality room and most, if not all, CPAs showed up. I had my file of REA correspondence with me and proceeded to show copies of the harassment I had been subjected to for years. It appeared that everyone in the room was encountering difficulty pleasing REA in their audit reports.

When they read some of the letters I wrote challenging REA, one of them said, "I would not have had the nerve to do it. REA is taking exception to my audit reports but the letters are mild compared to some directed at your reports." Before the shop broke up on Saturday, Andie Conners, CPA of Louisville, Kentucky, approached me and said, "Walt, I had planned to fly home tonight but I think you will need help tomorrow when REA meets with you and your Kansas clients. I am going to call my wife and postpone my flight." I expressed my sincere appreciation. Later that evening Andie and I talked to people with influence in Washington to determine how Andie could help me in a meeting called by REA. Andie really had no right to be at the meeting. Someone suggested that Andie make his way outside the meeting room as soon as the meeting convened and remain there until I called him to substantiate statements I was going to make at the meeting.

On Saturday evening I proceeded to a conference room just below Norman Clapp's suite to see what this prelude to the real action was all about. Chairman of the conference was Richard Dell, Deputy Administrator. John Scott, Leslie Surginer and James Sullivan of REA were present. The meeting took on the air of an informal court case. The plaintiffs were Scott & Surginer and the defendant was W.G. Schmidt. Richard Dell acted as judge and jury.

Through this relentless siege by government accountants it was my conclusion that the administrator or the area heads who signed many of the critical letters were mislead by their own employees. This was the approach I took with Mr. Dell. When my turn came to answer the charges leveled at me by the plaintiffs I unloaded ten years of frustration trying to please the banker. Copies of audit reports that had come under fire by the plaintiffs were handed to Scott and Surginer and I asked them to point out instances where I failed to meet either AICPA standards or the Minimum Standards of REA. Neither one took up the challenge. After the meeting I heard one of them ask, "Where do we go from here?" Mr. Dell pointed upstairs and I assumed he was pointing to the administrator's suite.

Sunday afternoon my Kansas clients arrived at a conference room at the Hadden Hall Hotel and so did Leslie Surginer of REA. After the usual amenities, Mr. Surginer presented REA's position reading from a manuscript on yellow legal paper. It praised the CPA audit program, with a few exceptions and avoided the acrimony of the past. I told my clients of my meeting with REA the evening before. When I came to my disclosure that I had evidence that REA discredited not only Walt Schmidt but other CPAs, Leslie Surginer rose to his feet. While he was trying to convince my clients there was no campaign against CPAs auditing REA borrowers, I walked to the door and outside the door sat Andie Conners pretending to read the daily paper. When Surginer finished his tirade, I introduced Andie Conners, explaining that Andie was a former REA field accountant and now had a successful CPA practice in Kentucky and Indiana much like my practice in Kansas. Andie addressed my clients, "Walt is telling you people the truth when he says other CPAs and their clients are getting the same critical letters. I receive them and so did the CPAs attending the workshop for CPAs just concluded." After the meeting Surginer accosted Andie and asked, "Why in the hell did you have to support him?" Andie replied, "It's simple. Walt was telling the truth." There were probably fifty people attending the Sunday afternoon meeting.

A larger meeting room was selected for the Monday meeting and about two hundred people attended. John Scott presented REA's statement by reading from a manuscript on yellow legal paper just as Surginer had done the day before. The states of Colorado, Wyoming, Nebraska, Iowa, Oklahoma, Missouri and Louisiana were present. My report included what had transpired at my meeting with REA on Saturday and the meeting with the Kansas clients on Sunday. After John finished his report, a man in the audience remarked, "And you pulled us out of an important NRECA meeting to tell us this."

At the request of REA, I agreed to alter my return trip and come home by way of Washington, D.C. where we accountants could discuss our differences under calmer conditions. In retrospect, it was another way for Scott and Surginer to continue their skulduggery.

While all of this was taking place, the remnants of a hurricane hit the east coast. Wave after wave of sea water came ashore flooding basements of hotels, knocking out electric power and demolishing parts of the board-walk. To my knowledge it did not flood the convention center but delegates had to walk from the top floors of high-rise hotels and wade through water to get to the convention. When I checked out of my motel the porter was losing the battle trying to keep the water out of the motel office. There were no flights out of Atlantic City. I was able to get a railroad ticket out of the city but the train was not able to back into the station due to high water. A high-wheeled bus was provided that took me and others to the edge of town where the train was parked waiting for customers from the depot. After climbing up the muddy bank, I was finally on a train for Washington, D.C.

A meeting in REA's office was held March 7, 1962 between W.G. Schmidt, CPA and members of REA staff, Leslie Surginer, Hubert Moore, Howard C. Paine and Clark Phillipi. A MEMORANDUM OF UNDERSTANDING was issued after the meeting signed by W.G. Schmidt and Leslie Surginer. It didn't admit anything and it didn't correct anything.

The parade of letters continued. A letter dated July 20, 1962 includes this:

> *After re-reading the sections of your audit reports relating to memberships of these borrowers, we agree your comments may be interpreted to say the cooperatives are complying generally with the provisions of their bylaws concerning memberships since no statement to the contrary is included in the report. However, where appropriate, we would like for you to make a more positive statement with respect to the bylaw compliance.*

> *We agree that your disclosure with respect to the numerical test of membership and service connection fees as compared with the number of meters in service, technically complies with our Memorandum of Understanding. However, here again, we would prefer a more complete statement indicating the technique employed in the test and your observations regarding differences that may be disclosed.*

On May 7, 1963 I notified REA that my practice was merging with Paul Schmidt as a partnership under the name of Schmidt and Company, CPAs. Mr Surginer's response to this follows:

> *We are sure that you are aware that when one CPA firm merges with another (or there are other changes in organization resulting in a change in the firm name), we have requested our borrowers to take formal action and submit a resolution to REA covering the selection of the firm under the new name if they wish to continue the auditing services. This in most cases is probably a formality; however it is important that there may be no question that the borrower and REA have approved the new firm to perform the audits.*

Nothing in my records indicates that REA requested such action on the part of our clients. I was the partner in charge of REA borrower audits so nothing changed.

It appears that from this day forward Leslie Surginer picked up the tomahawk and fought the battle alone. From 1952 to 1966 Surginer and cohorts were attacking my audit reports. The meeting in Atlantic City apparently put the damper on that approach. As mentioned previously, our audit agreement with the borrower provides that REA may review our audit work papers at a mutually agreeable time in our office. On August 1, 1966 this letter arrived from Surginer:

> *We have reviewed our travel schedule for the month of August and have found that arrangements can be made for Mr. Hubert Moore, Chief of the Audit Review Branch and Mr. Ernest F. Mercer, Supervisory Auditor, to meet with you during the week beginning August 29. They will plan to arrive in Kansas City August 28 and will contact you by telephone during the morning of August 29.*

As you might expect, on December 19, 1966 I received a letter from Mr. Surginer over two pages long reciting deficiencies in our work papers as discovered by Messrs. Moore and Mercer.

My letter in response is dated December 22, 1966:

> *Your letter of December 12, 1966 based on reports furnished you by your Messrs. Moore and Mercer came as no surprise. You should know and your borrowers do know that we do good work. Messrs. Moore and Mercer spent three or four days in Kansas City to visit two CPA offices. They looked at*

> *two or three work papers here and did not look at any work papers at the other CPA office. The information they asked for would hardly qualify them to pass the CPA test on audit standards or audit work papers.*

A letter from Mr. Surginer dated October 30, 1967 includes these comments:

> *When I visited with you last week at Missouri State Bookkeepers and Accountants annual meeting in St. Louis, Missouri, I mentioned to you our desire to make a further review of your work papers developed in connection with the audit of our borrowers' accounting records.*
>
> *Our travel plans for November show that Mr. Eugene R. Riddle, Chief Borrowers' Audit Standards Staff, and Mr. Ernest F. Mercer, Staff accountant, can meet with you during the week beginning November 13.*
>
> *If for any reason this date is not acceptable to you, please let us know.*

Another letter from Surginer dated November 6, 1967 states:

> *This will confirm our telephone conversation of last Friday, November 3, 1967 regarding the proposed visit to your office by Mr. Riddle and Mr. Mercer of our Audit-*

ing Standards staff. According to their present plans they will arrive early November 21 and will remain at least through November 22; if their work is not completed by this time, they may extend this visit through the holiday to November 24, 1967.

In accordance with our usual practice, arrangements are being made for one of our local field accountants to participate in the visit.

Your continued cooperation on this matter is appreciated.

A letter from Surginer dated February 21, 1968 covering over two pages states that his auditors substantiated all the deficiencies found in 1966 and discovered many more problems with our work papers.

My response to this letter contains the following:

In 1966, your staff visited our office and looked at two or three work papers. On that trip they visited another CPA office in Kansas City but no work papers were examined according to a telephone call from the CPA office. In 1968, your staff visited our office and looked at fifteen audit work papers. In your letter of February 21, 1968, you state 'as was the situation in August 1966, they were unable to determine from your working papers that your audits are being made in accordance with generally accepted auditing standards.'

> *Before planning a trip to Washington, I would like to know what 'evidence' or what 'documentation' would you as a CPA add to your work papers that Schmidt & Company has failed to include with its work papers?*

Surginer's letter dated April 23, 1968 contains this threat:

> *We still feel that such a meeting is desirable to resolve differences of opinion, to clarify our requirements and to establish a common course of action to assure that audits made of our borrowers' records meet our requirements. However, if you do not care to accept our invitation, we will have to pursue some other course of action.*

The siege against Schmidt & Company via its clients was only part of the strategy. State accountancy boards came into play also. One day I had a call from the Accountancy Board of Louisiana telling me that someone from the Department of Agriculture had alleged that Schmidt & Company was practicing in Louisiana without proper licenses. We explained to the accountancy board that we had no intention of maintaining an office in Louisiana and that we restricted our practice to our regular schedule of auditing REA borrowers. Our clients in Louisiana wrote letters to the accountancy board confirming our statement. The Accountancy Board of Louisiana awarded a CPA certificate to my partner and then wrote, "Now you can audit any one in the State of Louisiana." The Missouri Accountancy Board was also notified by Surginer that Schmidt & Company's audit work papers did not meet accepted standards. He provided the Board with a list of our clients and the audit fees charged to each client. In that case the Board asked me to meet with them and bring some of my work papers along. They were impressed with our preprinted audit procedure check lit, our internal control check lists and our proforma financial statements all designed especially for REA borrowers. When I asked what additional documentation I should add, the chairman of the board representing a national firm stated, "I can't speak for other firms but I think we accumulate too damn much paper. We are in paper up to our butts. In litigation, the lawyers kill us with our own papers."

In a letter from Surginer dated August 2, 1968 this announcement appears:

> *Since we have been unable to determine from your working papers that you are meeting the standards prescribed by REA, we are placing your firm on probation for one year beginning the date of this letter. During this year we shall not approve additional engagements of your firm to perform auditing services for REA borrowers. We regret that this step is necessary but we see no other way to discharge our responsibilities in connection with maintaining high standards in the REA program.*

The Kansas Managers Association also got into the act. While I was conducting the audit at Ellsworth, Kansas, Allen Paul, manager of the cooperative and chairman of the Managers Association was on the phone trying to get someone from REA to meet with our clients at NRECA's Region Seven District Meeting in Lincoln, Nebraska. It took the better part of two days before Allen cornered someone authorized to speak. REA finally agreed to such a meeting and I provided Allen a list of our clients and mailing addresses so he could invite them to the meeting. NRECA objected to such a gathering before their meeting but had no objection to our meeting after NRECA finished its program. I understood a room at the Cornhusker Hotel had been reserved for us and a time set for our meeting. Mr. Clapp flew in for the meeting and immediately invited Allen Paul to his suite. I have no idea what Mr. Clapp told Allen but after their meeting Allen washed his hands of our meeting. In fact he said, "There will be no meeting." To which I countered, "What do you mean there will be no meeting? You invited our clients and there will be a meeting." Luckily, the chairman of the meeting for the last day of the program represented our client at Grant, Nebraska. I explained my problem to him and asked if he would make an announcement over the public address system to the effect that Schmidt & Company would like to meet with its clients right after the meeting. He asked me to write my message on a card so he could read it. Shortly thereafter REA made a similar announcement over the PA system. Should any of our clients get to read this autobiography, I am certain some of them will recall the meeting. Incidentally, our reservation for the room had been cancelled but the hotel gratiously allowed us to remain in one of the meeting rooms. After our clients had gathered from Kansas, Nebraska, Wyoming and Colorado, Mr. Clapp and his entourage, including Mr. Surginer, came parading down the center of the room among the clients and presented their case from the floor.

The room was equipped with a stage, speakers' table and public address system but I was the only one at the speakers' table. Mr. Clapp took charge as if it was his meeting and when he and Mr. Surginer finished they started walking toward the exit. He must have assumed our clients would follow him to the door leaving me with an empty room. If that was his strategy, it backfired. There was no applause when Clapp concluded his comments and started shaking hands of acquaintances as he made his way through the group to the door. At that moment I took charge of the loud speaker and all of our clients remained seated. I recited my litany of harassment by REA and when I finished I received a standing applause. There was no longer any question but that Mr. Clapp was protecting his hatchet men.

We had another presidential election in 1968 and Nixon won so Dave Hamil was back as REA Administrator for another term.

I received this letter from the American Institute of Certified Public Accountants dated February 25, 1969:

> *Under the date of November 27, 1968, Norman Clapp, Administrator of Rural Electrification Administration asked the American Institute of Certified Public Accountants to make an independent evaluation of the audit work papers and procedures employed by you to determine their adequacy under the REA audit program. This request was made after much correspondence and discussion between the REA representatives and yourself.*
>
> *This federal program is of importance to the accounting profession as it has been a model program for many years and over five hundred firms are involved in audits. Accordingly, I accepted Mr. Clapp's request and appointed the following committee:*

Karney A. Brasfield, CPA, Chairman
Touche, Ross, Bailey & Smart
815 Connecticut Avenue, N.W.
Washington, D.C. 21006

Jamie L. Anderson, CPA
Main Lafrentz & Co.
P.O. Box 688
Waco, Texas 76703

James T. Malone, CPA
Stoy, Malone & Company
7315 Wisconsin Avenue
Washington, D.C. 21104

The committee has met with the REA administration to discuss this matter and has reviewed certain reports and copies of correspondence. However, the committee believes it of the utmost importance that it confer with you to become acquainted with all aspects of the problem, especially as to your viewpoint. Accordingly, the members of the committee would be pleased to meet with you at your office sometime in early March, if you agree.

May I please hear from you as soon as possible.

Very truly yours
Ralph E. Kent

Mr. Kent was president of the American Institute of CPAs at the time. I wrote Mr. Kent March 3, 1969 and told him I had called Mr. Brasfield and arranged a meeting for March 21. On March 24, 1969 I wrote this letter to Mr. Kent:

> *Mr. Brasfield and his committee met in our office Friday, March 21 to conduct an evaluation of our work papers and procedures employed in the conduct of audits for REA borrowers. The committee conducted its work in a very courteous and proficient manner. Working with them was a pleasure. My apologies to them for the limited quarters in which they had to work.*
>
> *At the conclusion of the evaluation, the only criticism voiced by Mr. Brasfield was with respect to 'Sufficiency of Evidential Matter' particularly as to the testing of samples. I have about worn out page 36 and 37 of AICPA Auditing Procedure #33 and underscored many lines. One of the problems involved in defining evidential matter is that CPAs, exercising individual judgment, seldom arrive at the same conclusion. Nevertheless, we discussed the subject at length and I assured the committee we would immediately add more footnotes, photocopies, adding machine tapes, etc. to our work papers, as evidence of tests performed.*

On May 15, 1969 Mr. Kent wrote this letter to me:

> *As agreed the Ad Hoc Committee on REA has independently evaluated the audit work papers relating to the examinations of REA associations conducted by your firm.*

We will forward a copy of the Ad Hoc Committee's report to Mr. Leslie Surginer in a few days but wanted you to have a copy in advance of the forwarding.

I appreciate your cooperation with the Institute in this matter.

The members of the committee appeared to be impressed with our work papers and audit procedures. One of them said, "This is the most specialized set of work papers and work procedures I have ever seen." It was manifest that Mr. Brasfield, the chairman, was not impressed and that attitude remained unchanged throughout the evaluation. A little research revealed that Mr. Brasfield's audit firm was conducting the audit of NRECA in Washington, D.C. Without any proof to substantiate my contention, I still suspect the American Institute of CPAs, in its anxiety to placate REA, allowed Norman Clapp or Surginer to name the chairman of that committee.

The Ad Hoc Committee report to which Mr. Kent refers in his letter of May 15, 1969 was terribly damaging to Schmidt & Company and might well have driven the final nails into the coffin. Upon receipt of the letter from Mr. Kent, I called him on the telephone and told him, "Mr. Kent, with due respect to the committee, many of the charges made in this report are not true and if you permit that report to go to REA, Surginer will kill us." Kent replied, "I was afraid of that. Can you be in my office Monday morning at 9:30? I will ask Mr. Brasfield to meet with us." I told him I would see him Monday morning. He agreed to reserve a room for me at one of the hotels for Sunday night near the AICPA office. If my memory serves me correctly, the hotel was one used by Arthur Young's partners and Mr. Kent was one of the partners.

This was my first visit to headquarters of the American Institute of CPAs and my reception was just as warm and courteous as my reception had been with my clients and friends in the midwest.

Mr. Brasfield arrived on time and the meeting opened on schedule. In both of the visits to our office by REA staff people to examine our work papers, I asked our secretary to provide them with any work papers requested but to keep an accurate record of who examined which work paper. The same procedure was followed with the Ad Hoc Committee and the only work paper Mr. Brasfield looked at was the electric cooperative audit at Bourbon, Missouri. Luckily, it was an audit I personally conducted from the engagement letter to submission of the audit report. My portfolio contained a copy of the work papers for the audit at Bourbon, Missouri. Mr. Kent opened the meeting by addressing Mr. Brasfield, "Mr. Schmidt takes exception to your statements in the committee report." As each of the exceptions to our work papers was addressed I would turn to Mr. Brasfield with the question, "Mr. Brasfield, you examined this set of work papers. I wish you would tell me and Mr. Kent what you would add to your work papers that I failed to include with mine?" I felt rather sorry for him because he evidently was in no position to judge what a good set of work papers for an REA borrower should contain. As we took each allegation, Mr. Kent would ask, "Mr. Brasfield would you agree to remove that item from your report?" and each time Brasfield consented.

After rewriting the committee report, Mr. Kent asked Mr. Brasfield to call the other members of his committee to see if they would agree to the new version of their report. The members agreed and so did I. Then Mr. Kent said, "Mr. Schmidt I am going to dictate a letter of transmitttal to REA and I want you to see if you have any objections before you leave here." The letter was satisfactory with me and I returned to Kansas City. On May 26, 1969 the committee report was mailed to REA.

On July 15, 1969 I addressed a letter to Mr. Hamil containing this information:

> *Mr. Weitzell, Mr. Brandts and I had a short, pleasant conference at Colorado Springs, June 24 regarding Mr. Surginer's allegations. Since Mr. Surginer failed to attend there was no opportunity to review our work papers with him. The cloud of suspicion hovers over our firm as it has for almost a year. Virtually every audit review letter to the borrowers questions the adequacy of our audits.*

Mr. Hamil replied on July 22, 1969 and stated:

> *Mr. Weitzell and I have read your letter of July 15 and note your concern about the status of your firm in conducting audits of REA borrowers. We sincerely regret that this matter has 'dragged' on for so long. It is our intention to reach a decision with respect to our relationship with your organization at an early date.*

During NRECA's meeting in New Orleans I had a chance to visit with Mr. Hamil. He told me, "We haven't forgotten your problem. Be patient, we have to do this in a certain manner. I am going to appoint another committee and the members of the committee will be unknown to you and Surginer."

A letter from E.C. Weitzell, Deputy Administrator dated September 22, 1969 contains these remarks:

> *We appreciate the cooperation Messrs. Bark, Heard and Johnson received during their recent visit to your office August 26-28, 1969. As we discussed earlier, the objective of this visit was to determine, through examination of working papers supporting your recent audit reports, the degree of progress your firm has made in meeting our audit and working paper requirements since our letter of August 2, 1968.*
>
> *In summary, our representatives found sufficient improvement in your procedure and working papers that we have decided to remove your probation, provided you will give us written assurance that you will further strengthen your procedures to eliminate the weaknesses noted above. We shall appreciate receiving these assurances at an early date.*

On October 7, 1969 E.C. Weitzell sent me the following letter:

> *In view of the improvements made in your audit program and procedures as evidenced by the working papers recently reviewed by our staff, and the assurances contained in your letter of September 26, 1969, that you will make further improvements so that REA auditing and reporting*

> *requirements will be fully met, we are today removing your firm from probationary status.*
>
> *We plan to continue to make periodic examinations of your audit procedures as reflected in your working papers and expect the next review to show that you are meeting all of our requirements.*

One of my friends in REA asked me one day, "Do you want to know what really took place? Mr. Hamil called in Mr. Weitzell and gave him this order, 'Call Mr. Surginer and tell him he has no case against Schmidt & Company and we are releasing them'."

On September 29, 1969 we mailed a bulletin to our clients including this statement:

> *Words are inadequate to express my sincere gratitude for the wholehearted support we received from so many REA borrowers during this most difficult year. It took courage to come to the aid of a CPA firm placed on probation. To my knowledge, we did not lose a single REA audit client attributable to the probation.*

By their deeds you shall know them. Schmidt & Company was too independent and had too large a following to please Mr. Surginer. To dislodge such a firm, a carefully calculated plan began to unfold. First the foundation was laid by letters from his audit review staff taking exception to our audit report followed by taking exception to our work papers. A bulky file of letters was accumulated of correspondence between Mr. Surginer and Schmidt & Company. Next Mr. Surginer sought more evidence by appointing some of his staff to committees to make an on-premise investigation to confirm the accusations made by the audit review staff at the beginning. Finally he was able to get the American Institute of Certified Public Accountants involved, when his case was falling apart. But for Mr. Hamil, a rancher from the midwest and a very popular administrator, he might have succeeded. The facts are that all of the charges of improper disclosure and inadequate work papers followed by letters claiming that Schmidt & Company made substantial improvements, were so much hog-wash. Schmidt & Company's audit standards and reporting standards have remained virtually the same during the entire ordeal. Our audit work papers are preprinted and were not changed. During an audit at Andarko, Oklahoma, I asked the manager, Elbert Karns, who was with REA for a time, "Look at this letter. Surginer is trying to destroy my audit practice. You were with REA and know how the system works. What can I do?" His advice was, "Walt, you will have to fight this thing or they will kill you." I took his advice and I saved my audit practice for the clients and for my employees.

As this biography states, most of my life has been involved with rural cooperatives in one form or another. Starting in the depression, it seemed like everybody was searching for the answer to the demise of so many cooperatives, grain, oil, mutual telephone, etc. As an auditor-analyst for CCA (now known as Farmland Industries) we would urge boards of directors to sell for cash and to limit borrowings to 25% of total assets. It was generally agreed that selling on credit and excessive borrowings spelled the death-knell of the cooperatives. We were constantly on guard to protect the exempt status of the cooperative with respect to the corporate federal income tax. These were also the desired objectives of the Wichita Bank for Cooperatives, the banker for many of the cooperatives. My thirty-two years of auditing REA borrowers, has convinced me that these objectives and this banker-borrower relationship does not apply to the REA program. Like any banker, REA insists on the borrower meeting the debt-service payments but unlike a banker, REA does not encourage the borrower to maintain a strong equity position. In fact REA encourages borrowers to apply for new loans as the old loans are fully paid. REA personnel have urged some of our clients to pay cash dividends to patrons when the members' equity was ten percent or less. A conscientious manager who operated conservatively and accumulated funds for emergencies was asked to dispose of these funds to qualify for a loan. The careless manager, who ignored operating in the black always qualified for a loan. I probably made no friends in REA with my advice to clients. I told them at public meetings, "In the long run this policy will cause you immeasurable problems. It may not be so bad when funds can be borrowed at two percent, but as the government interest cost increases, congress will raise the interest rate to you. Borrowing 90% or more of your capital needs, at the higher interest rates, will make it very difficult for you to compete with the utility serving a farmer across the road. Your problem of maintaining good relations with your members now, cannot compare to the public relations problems you will have, when your member has to pay more for a kilowatt hour of electricity purchased from his own utility than his neighbor pays buying from another utility." The day I predicted has arrived. In a number of instances the electric cooperative's retail rate is higher than the competing utility rates and public relations problems are mounting.

The area of exemption from income tax and my advice to

my clients also became a source of irritation to Surginer and Scott. Briefly, a cooperative pays no income tax because legally it is nonprofit. This means the bylaws of the corporation must demand that the corporation divest itself of margins by turning the margins back to the customer, like a sales discount. According to IRS this means calculating the amount and notifying the members of the amount. I doubt that many people working for REA have a cooperative background. My advice to our clients was to assign all margins to patrons and to notify the patrons as required by IRS. Surginer and Scott interpreted this as deliberate refusal to follow REA directives to the borrowers. Most of our clients followed our advice and that infuriated some people in REA. The cooperative at Sikeston, Missouri, followed REA's directive and left a portion of its profits or margins as unassigned or unallocated net earnings of the corporation. In our audit reports we simply disclosed what the cooperative was doing. REA did not use the terms common to accepted financial reporting such as "Retained earnings" but substituted its own terms "Operating margins" or "Nonoperating margins" followed by "Assigned margins" or Unassigned margins." The Sikeston client accumulated as "Nonoperating margins" sums for ten years and the aggregate became large. On one of my visits to Sikeston, I noticed some unusual activity in the accounting department. The office manager asked me, "Walt, do you remember when REA objected to your advice to allocate all margins? I received a request from REA stating, 'It is now our opinion that you should assign all your non-operating margins'. I had to dig out our patronage records for ten years, rent additional calculators and employ additional help to meet their demands."

There are nearly a thousand electric borrowers engaging a CPA for their annual audit. Most audits are conducted by a local CPA who audits perhaps one or at most a few REA borrowers. Schmidt & Company does more electric borrower audits than any CPA firm in the country. It was a lot easier for Surginer and Scott to impose their directives on a CPA that conducted one REA borrower audit. The fact that we had such a following, plus the fact that I had the audacity to correct some of the directives from Surginer and Scott probably contributed to the personal vendetta.

One does not forget a good friend and I didn't forget Andie Conners who came to my aid in the REA meetings of my clients in Atlantic City in March of 1962. My youngest daughter married a young man in the navy and moved to Norfolk, Virginia where Pat was stationed. While she was living there we planned a trip to the east coast, then north to Niagara Falls, through a portion of Canada and back through Chicago where most of my family resides. On the trip east we stopped at Louisville, Kentucky to see Andie. He was not in his office but some of his employees were on duty and told me Andie was working on his new corporation. At first I thought it might be a professional corporation but I discovered the new corporation was going public to raise money for a chain of fast-food restaurants similar to the Kentucky Fried Chicken chain headquartered in Louisville. He was starting a chain of "Andie's Catfish" drive-inns. Sometime later I called Andie from my office just to see how he was getting along and he told me, "Walt, I am in real trouble." So I offered, "Andie, one good turn deserves another. Is there anything I can do to help? You helped me once when in need." He commented, "You will have to come soon to help me." I flew to Louisville thinking he might have problems with REA but Andie's biggest problem was his chain of drive-inns. He was successful in raising about $500,000 to start according to his story, but construction costs took a lot of that capital and the one operating unit was losing about $2,000.00 per month. Andie was reaching retirement age and had sunk most of his savings into the venture. Like so many ventures, it failed. When I reached home I told my wife, "That darn catfish got Andie."

I met someone from Indiana later, who knew Andie, and I asked him about Andie's audit practice. I was told he was advised by REA to sell his practice to another CPA firm in Louisville. Was that Surginer's way of punishing Andie for supporting me in 1962? We may never know, but the possibility gives me a guilt complex.

Let me assure you, Walt Schmidt's problems with REA did not stem from his conservative nature and his dislike for federal government involvement. Just as the TVA in the southeast, Hoover dam in the west, Grand Coulee and Bonneville in the northwest, the electrification of rural America would never have been achieved by private risk capital. The Norris-Rayburn bill that created REA was one of the finest programs the Roosevelt administration brought to this country. When the law was enacted it was as a federal-self-liquidating loan program. The so called investor-owned utilities contend the program constitutes government ownership and operation of a competing utility industry. What the utilities neglect to tell the public is that government went to the industry and offered the loan program to them first but they wanted no part of a loan program involving the government. In fact they virtually declared war on the REA program with the result that REA and the local cooperatives reacted in kind. Did REA finance and construct power generation, when engineering studies concluded power delivered from the new generator would cost more than power available from an existing power source? Yes, there were such circumstances. On the other hand, did the power companies block generation for cooperatives until a loan was approved and then offer power at a lower rate? Yes, that took place. Did the utilities refuse to serve farmers until a loan was approved for a distribution cooperative and then rush construction of spite lines, stealing prospective customers from the new cooperative? Yes, the utilities are guilty as charged.

The final result is that the program brought electricity to people in the far corners of rural America and raised the standard of living for millions. It has also brought economic benefits to the nation beyond the most optimistic forecasts. Has the program been operated on self-liquidating loan basis? No, the program is celebrating its golden jubilee but in reality through its revolving loan program and its guaranteed loan program through National Rural Utility Finance Corporation and the Federal Finance Bank, government is very much the prime investor in the rural electric program. The cooperatives are in such poor financial condition that withdrawal of all government support at this time would result in disaster for the local cooperative in my opinion.

Let us dwell a little on the development of our REA audit practice from the beginning with one borrower audit in 1952 to the time when five CPAs in Schmidt & Company spent full time on those audits. What factors contributed to such growth? Three come to mind, my cooperative background dating back to the depression years, the quality of our audits and our audit reports and the efficiency of our practice. We were able to maintain competitive audit fees through the cooperation of the clients. Some clients have a twelve-month audit as of December 31, some as of November 30, some as of October 31, etc. When a borrower asked for an audit proposal we would ask the borrower for its present audit date and if it did not fit into our audit schedule we would insist on the borrower changing its audit date. By so doing, Schmidt & Company had no non-productive payroll. Substantially all accounting firms pay CPAs for hours that cannot be billed to the client. The rates to the client have to be increased for the non-productive time.

Schmidt & Company had a policy of attending the NRECA Region VII meeting every year and Region VIII meeting at intervals to keep in touch with changes in the program. Schmidt & Company accepted invitations to appear on statewide REA accountants programs at no cost to these groups. Publications were ordered from surrounding REA statewide offices, as well as, the national office of NRECA. One of the first accountants' meetings I attended was as Scottsbluff, Nebraska and an employee from the McCook Public Power District approached me and asked, "Our manager wants me to ask you whether it would be possible for you to stop at McCook on your return to Kansas City?" I told her that was no problem. My stopping at McCook brought our first Nebraska client. Sometime later Pappy King, manager at Axtell, Kansas, asked, "Do you know why you got the audit at McCook?" I told him, "I assumed one of you guys in Kansas recommended us." Pappy added, "It happened at Cody, Wyoming at a regional meeting. I and Fred Atkins from Solomon and George Myers from McCook were sharing a bottle in my room and either Fred or I asked George, 'Who do you plan to hire for your CPA audit?' George responded 'Probably Jack Tarr with Peat, Marwick, Mitchell because I know Jack.' Before the night was over we convinced George we had the CPA he needed."

My first client in Missouri at Savannah resulted from a visit at my home by the manager and his secretary. My first clients in Wyoming came through a phone call. Emil Blum, manager at Saratoga called and told me some of the managers were unhappy with their audit connections and wanted to know if we were interested. I was reluctant to travel that far but I told him, "If you can get a couple of manager to come to Wichita for the regional meeting I will meet with you to see what can be done." As a result of that meeting my first trip to Wyoming was for audits at Lusk, Saratoga and Wheatland, Wyoming, in a bright red new 1956 Pontiac two-door hardtop. Our first stop was Mitchell, South Dakota, where we visited the corn palace in early June. The car did not have air conditioning so we drove with windows down. The second day was hot with strong south winds blowing through the car and three pigtails flying in the back seat. That afternoon we passed through the Black Hills with its famous presidential monument. We found space in an older motel in Lusk with cooking facilities so Louise could keep some fruit juice and snacks at the motel for our three daughters. I am certain they remember feeding raw carrots to a white horse in a pen near our motel.

In Saratoga, the manager's son shared his pony with the girls and they took turns riding it. When the audits at Thermopolis and Basin, Wyoming were placed on our schedule, the girls got to see the Yellowstone National Park, the Teton Range and the Mormon Cathedral in Salt Lake, Utah. On our way home we visited the dinosaur museum in eastern Utah.

We have a daughter living in Kansas City and in later years she and her two daughters accompanied us on a trip to northwest Wyoming and we took the grandchildren through the park. When we left the Holiday Inn motel in Thermopolis I asked the girls, "Where would you like to stay tonight, in a modern motel or would you like to rough it in Teddy Roosevelt's Hunting Lodge?" Both girls voted to stay in the hunting lodge at the north end of the park. The lodge had a wood burning stove, electric lights but not modern plumbing. For those facilities one had to walk to a central location. In early summer the nights are quite cold in that part of Wyoming and it was cold that night. There was a restaurant on the premises so we were able to get breakfast at the lodge. After breakfast we continued our leisurely tour through the park. When we reached the Old Faithful geyser, I spoke with the girls about lodging for the night. This time they voted for more modern facilities.

One year we took a short vacation at Estes Park, Colorado and stayed at the YMCA park. We had a nice cabin finished in southern yellow pine, and the summer programs for the children were in full swing. It was there that I noticed the name of a classmate in Hays high school. Orvis Grout, known as Gravy Grout by his classmates, was the son of a Watkins Products distributor that came by our farm at intervals. Gravy had a part time job at a local barber shop in Hays. In class, Gravy enjoyed nothing better than to entertain his classmates to the distraction of the teacher. Not only did he forget his daily assignment, he frequently left his text at the barber shop. There was one assignment that Gravy respected and performed superbly and that was performing on stage. I graduated with Gravy in 1930 and had not heard from him since. At the YMCA camp, I picked up a program and on the front of it was the announcement that Dr. Grout, Doctor of Drama, Greeley State College, was in charge of the program.

Some of our favorite locations in the mountains were Steamboat Springs, Vale, Gunnison, Ouray and Durango, Colorado. During one of my audits, our three daughters and their children joined us for a week of relaxation. It was a week all of us will remember for a long time. I had reserved space on a Denver Rio Grand narrow gauge railroad that tourists take up the canyon from Durango to Silverton. We rode the train up the valley to Silverton and then took a bus over the mountains coming back. A very scenic and beautiful trip. One day Louise and the daughters planned a program of their own and I agreed to teach the grandchildren how to fish. Distribution of the fishing tackle to four youngsters was a lot easier than retrieving the tackle. They hooked everything from rocks and tree limbs to each other. We finally compromised and fished at a trout farm where grandpa financed two trout for each grandchild. That evening we had a fine picnic at a little park along the river with roasting ears and broiled trout. One day was devoted to a tour of the Mesa Verde National Park.

In a prior trip to Durango my oldest daughter and her two daughters were with us and Karen pulled a toe nail while playing on some school ground equipment. We rushed her to the hospital for emergency care and to our surprise the doctor's name was Pivonka, my wife's maiden name. Louise and the doctor discovered no blood relationship between the Pivonkas. During our stay with the grandchildren on this trip, the youngest grandchild, Eric a toddler, lost his balance and bumped his head on the corner of the fireplace at the motel. We rushed him to the emergency room to repair a break in the skin on his forehead and to our surprise the same Dr. Pivonka mended Eric's forehead. When he finished he turned to me and said, "Grandpa, you don't take very good care of your grandchildren."

Louise usually traveled with me to Steamboat Springs, Colorado. This audit was conducted in February when the grandchildren were in school. It was one of the few Colorado audits performed on a calendar year basis. Steamboat is a famous ski area and a beautiful place in the winter. One year while we were there, Steamboat had its winter carnival with horse races down main street and night ski runs down the mountain side with each skier carrying a lighted torch. Some years, snow would accumulate to ten feet deep near our motel. Electric lights glistened off the snow on the mountain sides.

My audit practice entailed a lot of travel but most audits could be completed in the field within a week by working long hours so I was home on most weekends. The audits in Wyoming and Colorado were usually scheduled in pairs so the family would travel with me for two or three weeks. Louise also traveled with me on trips to various annual meetings.

According to AICPA's membership, about 25% of the CPAs in public accounting are practicing on an individual basis. Substantially all of the remainder practice on a partnership basis. In recent years some of the firms have formed professional corporations. As an individual practitioner, one secretary and not more than three non-CPAs constituted my staff. There was a turnover in secretaries after about the first year otherwise my first secretary and my first professional remained on my payroll for twenty-five years. Due to the heavy travel requirement of the practice, it was never easy to find competent accountants to audit the REA borrower but eventually competent accountants came aboard. One of the attractions was an employment agreement in which I shared revenue with my audit staff. It was my responsibility to see that revenue was distributed equitably among the audit staff. As an individual practitioner, the CPA is master of his practice and can devote full time to his clients. In a partnership or professional corporation, the organization and personnel problems can absorb a lot of the CPA's time. On the other hand, the individual CPA is handicapped in his attempt to attract larger audit clients. The REA borrower for example prepares budgets and financial plans five or ten years in the future and is concerned about the dependability of its consultants such as attorneys, engineers and CPAs. In other words what happens to their CPA connection if Walt Schmidt should die or become permanently disabled? Continuity of the practice is important to the professional staff as it provides hope for advancement. I practiced as an individual CPA for eleven years and one statistic troubled me greatly. So many CPAs developed a sound practice over a lifetime and then watched it disappear because the CPA failed to provide for continuity of his practice. I was determined this would not happen to Walt Schmidt and his clients.

Like a marriage, however, it has to be the right partner. As mentioned, I found such a partner and in addition to the organization of Schmidt & Company CPAs, a partnership, Paul Schmidt and I formed Schmidt Investment Co. Inc., a corporation, that owned our office building. While Paul and I were working on these organization, Paul invited me to come to his office on 3611 Main Street. Louise accompanied me as we entered the office. As I recall, it was a large room with portable wood and glass dividers to form small enclosures for the accounting staff. The secretary was seated next to the door behind a cloud of smoke. While I smoke, my wife is a non-smoker and this was offensive to her. Paul has a good personality and made us feel welcome. When we left Paul's office to return to my office Louise remarked, "A partnership between you and Paul will never work. You are Mr. Clean and did you notice the papers scattered over the desks and piled up against the glass partitions in those offices?" I too had noticed the office was untidy and I disliked the smoking by the receptionist but I allowed for the very limited space.

The partnership did work for twenty-one years but Louise was correct, the untidy work habits of Paul and his staff created some problems in our new office. As office manager, I reserved the right and assumed the obligation of making our office look like a successful CPA office. No papers were to be left on the desks at night so the janitor could clean the glass and formica tops of desks. Shortly after we opened, the janitor came to me and asked, "What shall I do about Paul's office? He has papers strung all over." I told him, "I don't like it either but after all Paul is a partner. Just empty his waste basket and do the best you can to clean his room." About half of Schmidt & Company's professional staff was on Paul's payroll and when they saw how Paul honored my instructions on office appearance, they were reluctant to follow orders from me. In fact one of his employees left after he and I confronted each other one evening. He was ready to leave the office with a lot of papers on his desk and when I asked him to remove them so the janitor could perform his assigned task, he rebelled. I told him if he didn't remove them before he left, I would pitch them into the waste basket and the janitor knew what to do with paper in the waste basket. I had assumed that all CPAs were neat and orderly in their appearance and in their work habits. Paul's untidiness bothered me but it did not wreck our partnership. Paul was neat in appearance but no one could accuse him of being neat in his work habits. I would ask our secretary, "How in the world does he find anything in that mess?" She would say, "Don't ask me, but he does." Instructions from the office manager included such things as no coffee cups on desks and no smoking at desks. Coffee was furnished by the firm but it had to be consumed in the coffee room and smoking was confined to the same room. Attire could be in the form of jeans but dresses were preferred. We retained Paul's secretary as our receptionist but when she learned of the smoking rule she found other employment.

While both of our ancestors came from Russia, Paul's were menonite people and came from the Ukraine. Their beliefs and lifestyle were quite different. One of the questions

Paul posed early in our partnership discussion was my attitude toward conscientious objectors. Paul served in World War II but did not bear arms. Our ideas of what a partner's office should look like were certainly different. While I was a working partner, I saw no need to accumulate papers around the walls of my office to impress the client. In fact it did not impress a group of REA managers when I gave them a tour of our office and introduced them to Paul. Our offices were side by side and after seeing Paul's office one of them said, "It's not hard to see who's boss around here."

The partnership agreement in effect when I retired provided for senior and junior partners. Our first partnership agreement was executed when I was fifty-three years old and Paul was about ten years younger. One of the objectives of a partnership agreement is to provide for the orderly retirement of the founding partners. I prepared the partnership agreements using suggestions published by AICPA and obtained approval by Paul. As Schmidt & Company grew I tried to recruit professional staff that had passed the CPA examination. This was to lend credibility to the audits and to provide a successor to me in the event of my retirement. This became part of the employment agreement with CPAs but only one CPA would succeed me as senior partner and that would be decided when I retire.

Both of our practices were growing and so was our staff. The secretary was put in charge of two audit typists and was responsible for office supplies. She also hired and fired typists subject to my approval. My orders were that we would follow established lines of authority and responsibility. If Paul's people needed secretarial work it should be taken to the secretary and not to the typists. This was simply what I had learned as a manager and the result of some management courses I took while I was with Farmland (CCA). While Paul did not challenge my arrangement he certainly preferred the loose family style of office procedure where no one worries about lines of authority or responsibility. One wonders what would have happened to our partnership had I not insisted on the rule, "No employee of Schmidt & Company should be asked to serve two masters." From the first professional to the last, I had written employment agreements, whether he had his CPA certificate or not. To my knowledge Paul never had a written employment agreement with his professional staff.

Schmidt & Company was and remains a two member partnership. Senior partners control the partnership and provide its capital. The title "Junior Partner" is used in the employment agreement with CPA professionals to denote the fact that they qualify to succeed the senior partners. They provide no capital and exercise no control over the operations of the practice. My files indicate that Paul and I were corresponding about the retirement provisions as early as 1971. Retirement for a founding senior partner according to the agreement was optional after the age of sixty-five years. Successor senior partners shall retire at sixty-five years of age. I notified my founding senior partner of my desire to begin my retirement on July 1, 1977, to become effective December 31, 1977, and my retirement would culminate December 31, 1984. Subsequent to the effective date, the provisions for the retirement of the founding partner's capital and payment of his pension went into effect.

After I notified Paul of my desire to retire, my first professional employee Bill Hamlett submitted his resignation. A little later our secretary for the past twenty-five years, Marian Bayless submitted her resignation. Marian had been divorced for quite sometime and Bill became divorced about the time of his resignation. Before the year expired the two were married. Bill ignored the non-competition clause in our employment agreement, proceeded to solicit Schmidt & Company clients for a competing CPA and filed suit against Schmidt & Company for failing to pay him overtime. I countersued for breach of contract and asked the court to grant an injunction to prevent Bill from serving our clients. The case never went to court due to manipulations of the attorneys. Bill kept about six of our clients and apparently paid all legal and court costs because Schmidt & Company paid nothing.

Paul hired a young CPA by the name of Vaughn Cornish who came to us from a firm in Florida. Effective January 1, 1978 Paul entered into an agreement entitled "Partnership Admission Agreement" but in the first paragraph the agreement stated, "for the purpose of admitting Vaughn as a junior partner of Schmidt & Company." The agreement went on specifying "Junior Partner's Capital" which provided for payment of capital and "Division of Profits and Losses, Termination of agreement, and Settlement in the event of termination." This one act by Paul created a partnership within a partnership. Nothing in my partnership agreement with Paul bound me to a partnership with an employee of Paul's. A similar agreement was executed with another young man just out of college by the name of Kenneth Smith on January 1, 1983 after he passed his CPA examination while an employee of Schmidt & Company. These actions became a cancer eating away at the good relation Paul and I had for fifteen years. Vaughn expected me to recognize him as my partner, when in my judgment he was an employee. To complicate matters, Paul allowed Vaughn to become a distributor of Amway Products while he had space in our office and remained on our payroll. When he billed us for Amway products I paid the first invoice but told him not to present another. When he brought gratuities to our secretary and typists I protested to Paul about the conduct of his employee.

On November 17, 1982 I wrote a memo to Paul containing these statements:

> *The next two years will be very important to Schmidt & Company as a partnership. The test will be the transition from one of the founding partners to his successor.*
>
> *Our last payment on the building falls due June 1, 1983 and the investment company will accumulate some cash after that date. It would be my suggestion that the money be used to pay liabilities to partners, etc. until December 31, 1984 when the lease to Schmidt & Company terminates. It is to your benefit and mine that Schmidt & Company continue in practice just as it is to the benefit of Schmidt Investment Co., Inc. as landlord for the partnership. I would like to suggest that one of us acquire the other's stock at an agreed value, to be paid over ten years at the current interest rate on ten-year treasury securities, by offering as collateral a deed of trust on the land and building. At the same time Schmidt Investment Co., Inc. would execute a ten-year lease agreement with Schmidt & Company at a reasonable rent commensurate with the investment in the land and building. Both of these should take place January 1, 1985.*

On July 21, 1983 these comments were included in my memo to Paul:

> *Some weeks ago I mentioned to you that word came to me that Vaughn was signing partnership checks in your behalf and our agreement provided that Senior Partners be authorized to draw on partnership bank accounts. Instead of removing Vaughn's signature from your partnership bank account you propose to add Ken's name. I cannot in good conscience add my name to this authorization for the following reasons:*
>
> > *There is no need to have so many authorized signatures.*
> >
> > *If your junior partners have access to your checking account, is it not reasonable that my junior partners should have access to Schmidt & Company's bank account?*
> >
> > *It appears to me that you are attempting to convert Schmidt & Company from a two-partner firm to one with seven partners.*
> >
> > *My juniors and I have never discussed changing the partnership agreement for more than two general or senior partners.*
> >
> > *Senior partners may come and go. Schmidt & Company should continue to grow in revenue and reputation.*

My insistence on a two-partner firm came from advice I received when I resigned from Farmland. Shortly before I left, the company had employed Bob Norwood, CPA, who had been a partner in a CPA firm in Oklahoma City. Bob's job at Farmland was to take care of the company's tax returns. Bob told me, "Walt whatever you do, keep your firm small and manageable. I opened my office in Oklahoma City, rented an office in a high-rise building in downtown, bought some leather upholstered chairs and put my name on the window. The office looked fine but the clients didn't respond. Other CPAs in my position and I reached the conclusion at a local chapter meeting to present a better image we should form a partnership and include four or five names in the firm name and print that on the window. It was a big mistake. We could not agree on what book to buy for the library."

About the only response I received from Paul was that in his practice he needed a second signer on checks. "The partnership agreement specifies that you should be on it, but does not limit it to you. I have sent in the signature documents without you signature." Both Paul's bank account and mine for the partnership were with the United Missouri Bank but he used the bank downtown and I used the one in the south part of the city. My bank said their policy required a resolution signed by all general partners to authorize signatures.

Despite my objections, Paul reported his payments to Vaughn and Ken as partners instead of employees. Vaughn apparently concluded there was no way for him to gain control of Schmidt & Company so he concentrated on Paul's half of the partnership. In a memo dated October 31, 1983 Paul stated:

> *As I visualize this arrangement, we would incorporate not later than 12-31-83*

> *and the corporation would become a partner of Schmidt & Company. All fees would be billed and collected through the partnership and be withdrawn periodically by the corporation. Overhead would be reimbursed to the partnership (similar to the present method) except the corporation would take the deduction. Salaries for my area would be paid by the corporation. Client reports would continue to be under partnership name.*

My reply to his suggestion contained these statements:

> *Continuing the partnership under the amended partnership agreement dated November 1, 1983 is by far the best assurance that you and your estate will be reimbursed for the value of your practice and the capital you have in the practice at the time of retirement. Tax shelters should not interfere with a fine partnership practice.*
>
> *I have never heard of a two-partner CPA firm in which one is a CPA and the other is a corporation. Under no circumstances will I sign a partnership agreement with a corporation. Roger tells me that neither will he.*

In the statement above, I am referring to Roger Schmidt, CPA, who I selected to succeed me upon retirement. Roger is not related to me and had been on my staff for about fifteen years and was well received by the clients.

The memos in my file show clearly that the "Amway Peddler" had convinced Paul the pension benefit payments prescribed by the partnership agreement would constitute earnings from self-employment and therefore deny Paul the Federal Social Security Benefit payments. I wrote Paul a memo giving seven reasons why Schmidt & Company's pension payments did not constitute earnings from self-employment but he replied, "Barring unforseen circumstances we plan to incorporate as a professional corporation as of December 31, 1983 principally to prepare for my retirement when and if it arrives."

On November 21, 1983 my memo to Paul includes these statements:

> *Last Friday I spent most of the day preparing a lengthy response to your ultimatum of November 18, 1983. After sleeping over the matter I realized this exchange of memos is just an exercise in futility. Apparently you and your junior partners reached the conclusion to incorporate even if it resulted in the liquidation of the partnership. Contending that retirement under the partnership plan would deprive you of your social security benefits, is just camouflage for the real objective which is to dissolve the partnership between you and I. I retired under the partnership plan in 1977 and my social security check has been paid in full and on time ever since.*
>
> *I regard your threat to become a professional corporation as of December 31,*

1983 as your notice to me that you wish to withdraw from the partnership as of that date.

Our practice will continue as a partnership and yours as a professional corporation, completely separate from each other.

I contacted the Secretary of the State of Missouri and obtained a copy of the certificate of Incorporation of Schmidt, Cornish & Smith, CPAs, P.C. dated January 9, 1984. About a week later I asked the Missouri Division of Employment Security to retain possession of Schmidt & Company's unemployment compensation funds. I also notified the United Missouri Bank. Paul was wrong in leading his CPAs to believe they were partners of Schmidt & Company. An even more damaging blunder was when he allowed them to assume control of his practice. Under the partnership agreement, Vaughn and Ken had no choice but to honor their obligation to pay both capital benefits and pension benefits to Paul in the event of retirement. Under the corporate structure neither is obligated to make the payments and the professional corporation is not required to provide a pension fund. According to Paul, it is an unfunded pension plan, probably not worth the paper it is written on.

We exchanged memos and held meetings but were getting nowhere. Paul and his juniors insisted on forcing their corporation into the partnership and I was determined that would never happen. Besides relieving themselves of the obligations to Paul in the event of retirement or death, they also wanted their names in the corporate title.

It became clear to me that Paul and I had reached the end as partners of Schmidt & Company. When I asked Paul, "How much will you take for your stock in Schmidt Investment Co., Inc.?" he replied, "My stock is not for sale." I finally turned the whole matter over to my attorney. He wrote Paul a letter dated February 15, 1984 regarding both the investment company and the partnership in which he wrote:

> *"Subsequent to our being consulted by Mr. Walter Schmidt, I have reviewed the agreement under which you are now operating and wish to advise that there is a legal procedure for the dissolution when an impasse has been reached. Similarly there is a law for curing a deadlock in corporations which prevent sale of assets, etc. While he does not want to be involved in a court matter anymore than you should want to, he has instructed us to take that course. Hopefully, it will not be necessary. I would propose that you consider the above, and if need be, consult with your attorney along the lines that I have indicated above and advise me of your intentions within thirty (30) days so that I may act accordingly."*

The letter broke the ice and Paul made an offer for my stock in the amount of $50,000 but in the memo he stated, "I would not be interested in selling stock in the investment company until my interest in the accounting practice is terminated." He later suggested that a down payment of $10,000 in cash be part of the arrangement and I agreed. My offer for his stock was $60,000, his next offer for mine was $62,000, my next offer was $65,000, his counter-offer $66,000, my next and what became the cost of the stock was for $70,000. Schmidt Investment Co., Inc. paid Paul and Cathryn $10,000 in cash and delivered to them a ten-year note secured by a deed of trust in the amount of $60,000 as the shares of stock were surrendered to the company. The value of the stock was a guessing game because neither one of us was certain as to the market value of the land and buildings owned by the corporation. Paul may have relied on a real estate appraisal.

Our next encounter was to resolve our problem with regard to the conflict between his practice as a professional corporation and my practice as a partnership. The lease Schmidt & Company had with the investment company expired December 31, 1984. On April 1, 1984 I submitted a proposal to Paul attempting to achieve a working relationship between the professional corporation and the partnership so neither would have to move. I suggested renewing the lease with the partnership for $1,500 per month provided the professional corporation continued paying sixty percent of the rent as it had in the past. Office overhead would continue to be allocated based on a detailed analysis of the prior year's expense. The demands of the professional corporation were unbelievable so there was no need to consider renewing the lease. On April 20, 1984 I advised Paul that the office lease to Schmidt & Company would not be renewed as of December 31, 1984. During 1984 I was paying the overhead expenses and billing Paul for his portion. The last reimbursement I had from Paul was dated September 17, 1984 and was to apply on the August 1984 overhead. In a memo of May 7, 1984 to Paul I state, "It appears to me there is very little left to negotiate between you and I. Neither you nor I will be partners of Schmidt & Company in 1985. However, Schmidt & Company will continue to practice as a partnership in the future under the same name and telephone number until a permanent location has been found."

In our relationship it was agreed at the beginning that I would maintain the books for the partnership, since I was office manager, and he would maintain the records for the investment company. Since Paul's income tax practice was extensive, I suggested that he prepare the partnership income tax forms after combining our financial statements. During 1984 my file is full of Paul's memos accusing me of not reporting the capital and pension payments to myself properly for income tax purposes. The ordeal must have reached the limits of my patience as the following note to myself discloses:

> *Paul came to my office this morning and expressed his indignation at the fact that I did not answer his memos. I told him to fight with Vaughn Cornish and not bother me. I told him I resented his cross-examining me as if he were a prosecuting attorney. All the information he has been bugging me for has been explained in writing for more than a year.*

Our secretary called my attention to the fact that some of our engraved letterheads appeared to have vanished. A week or so later a delivery was made from a local print shop and it contained printed announcements on our letterheads as follows:

> ANNOUNCEMENT—WE ARE MOVING!!
>
> Schmidt & Company, CPAs is opening a new office. The following partners and

their staff are moving on October 19, 1984:

Paul Schmidt, CPA
Vaughn P. Cornish, CPA
Kenneth R. Smith, CPA

New Office:	Telephone
4420 Madison, Suite 170	(816) 753-6063

After these were mailed to Schmidt & Company local clients, I prepared this notice and mailed it to the same clients:

NOTICE

> Paul Schmidt, Vaughn Cornish and Kenneth Smith have withdrawn from the partnership to practice as a professional corporation. Articles of incorporation were filed with the Missouri Secretary of State January 9, 1984 for Schmidt, Cornish and Smith, CPA, P.C.
>
> The incorporators are in the process of relocating. Interested parties should note that the corporation represents a distinct separate legal entity and has nothing in common with Schmidt & Company.

My figures show when Paul left he owed me unreimbursed office overhead expense in the amount of $6,203.02. Instead of receiving payment, I began receiving letters from his attorney claiming the allocation of overhead expense, to which he had agreed, was inequitable. Schmidt & Company had an unemployment compensation fund balance of $9,672.61 and according to my calculations Paul paid in about

46% of the contributions. In other words, it appeared he was entitled to $4,449.40 of that fund. The Division of Employment Security ruled, however, "Since the corporation only acquired a portion of the business of the partnership, they cannot succeed to the account of the partnership under the present law governing successorship."

I threatened to sue Paul for the $6,203.02 owed but when the amount was reduced to $1,753.62, due to the unemployment compensation fund, any gain from a law suit became more questionable. I doubt that the $1,753.62 will begin to solve Paul's problems with his professional corporation.

When the professional corporation moved in October of 1984, I contacted the real estate brokers in Kansas City listing my priorities as follows: One, lease of the lower level of the building; Two, lease the entire building; Three, sell the building at 2837 Main. When Paul moved out he ceased paying rent so Schmidt Investment Co., Inc. lost sixty percent of its rental income. My portion of Schmidt & Company could not afford paying rent on the entire building. At first I sought another CPA in practice who might want to merge with us just as Paul had done years ago. By so doing my successors would not have to move. There was some interest but the search was fruitless. Leasing the lower level to another tenant was another alternative that did not materialize.

The time had arrived for me to pass control of Schmidt & Company to my successor as managing partner. Roger Schmidt who had been my choice as managing partner needed to select a partner and we needed to revise the partnership agreement to fit the new personnel and circumstances. New employment agreements were drafted for employees (Junior Partners) of the firm. Since the partnership accounting was kept separate between Paul and I, dividing our assets, obligation and clients presented no problem. Transferring my portion legally to a new partnership consisting of Roger Schmidt and Wayne Middlebrooks, senior partners, was accomplished by a purchase and sales agreement plus a bill of sale for office equipment. My understanding with my junior partners was that if they remained loyal to Schmidt & Company until December 31, 1984, I would turn the practice over to them. The price my successor senior partners paid for the practice including the library and equipment was "One dollar and other valuable considerations." For the accounts receivable at December 31, 1984, Roger and Wayne signed a note payable to me in the amount of $39,000.00 payable as follows: Principal payment of $19,000.00 on July 1, 1985 and $500.00 on the first of each month thereafter until the principal is paid in full. Interest was to be paid quarterly at Commerce Bank's prime rate.

It appeared to me that when control of Schmidt & Company passed from me to Roger, everyone on the payroll descended on Roger with suggestions for changes in the firm's policies. This is probably to be expected when authority passes from one individual to another. As managing partner I granted considerable leeway to junior partners but I did not hesitate to change their reports when I deemed it necessary. I reserved the right to assign or not assign an audit to a specific professional. Auditors were rotated and consideration was given to equitably apportion the revenue as well as the miles traveled by the auditors. This type of management was my advice to Roger and to assist him I offered to remain available as a consultant. When I saw how these people were influencing Roger's decisions, I warned, "My offer to stay here for a while as a consultant is contingent upon Schmidt & Company remaining consistent with its present policy. The day that Wayne and Bill dictate policy for Schmidt & Company, is the day I will put plenty distance between myself and Schmidt & Company." His response to my offer was, "The only way I can show those guys that I am in command is that you give up your office and let me have command of the business." He also inferred that a managing partner with such control was akin to a dictator.

My next memo to Roger contained these statements, "Before receiving your message, I sincerely felt obligated to do what I could to help you in your new responsibilities. This is to advise that Schmidt Investment Co., Inc. does not want a long-term lease with Schmidt & Company. You may remain in the building for the time being by paying $1,000.00 per month and paying the utilities. You will have to move when a suitable tenant is found or when the building is sold."

On February 6, 1985 my memo to Roger contained these statements, "Schmidt Investment Co., Inc. has this day accepted an offer for the sale of the real estate located at 2837 Main Street, Kansas City, Missouri. As stated in my memo of 1-28-85, Schmidt & Company may remain in the building for the time being by paying $1,000.00 per month and paying the utility bills. The rent should be paid in advance. Unless you can reach agreement with Hallmark Cards, Inc. for lease of the building, you must vacate same before April 30, 1985.".

There was little interest in leasing but two buyers became interested and asked me what I wanted for the building. I in turn asked them to make me an offer. Both parties came to the building several times to look at the facilities. An offer was finally made by one and then I made an offer to sell for $160,000.00. Both kept raising their bids and Hallmark Cards, Inc. finally purchased the building at my price. With that market value, Schmidt Investment Co., Inc. had a net worth at October 1, 1984 of about $165,000 the date on which the company purchased Paul's stock. Hallmark Cards, Inc. wanted to pay cash but I insisted on taking part of the principal in the form of a note, secured by a deed of trust, bearing interest at 12 percent. Paul's note was paid in full at the time of sale.

Schmidt & Company moved its office to Blue Springs, Missouri, a suburb of Kansas City during the month of March, 1985. Roger may not approve of my management of Schmidt & Company but he cannot deny that under such management my practice grew and prospered for thirty-two years. In my opinion a professional practice, like a business enterprise, functions successfully when control is vested in a manager. Properly constituted authority is absolutely essential. By functioning as managing partner of Schmidt & Company, I maintained control of my practice. Paul operated on a pluralistic basis and lost control of his.

As my successor, Roger Schmidt inherited a financially sound successful practice. Through the partnership agreement and the employment agreements, he has the necessary authority for good management. If Roger retains the respect of the clients and the profession, if he keeps a watchful eye for mischievous power-hungry bureaucrats, if he maintains top quality audits and if he leads rather than follows his subordinates, Schmidt & Company, Certified Public Accountants, should be around for many years to come.

On April 30, 1985 I turned the keys to the office building at 2837 Main over to Hallmark Cards, Inc. and moved my desk to the family room of our new home. As of this writing I am President of Schmidt Investment Co., Inc. and have a permit to practice as a Certified Public Accountant in the State of Missouri. I am a member of the American Institute of CPAs and the Missouri Society of CPAs. □

CHAPTER XI

Travels

With the beginning of my retirement as a senior partner of Schmidt & Company, CPAs, my wife and I decided to do some traveling. Since Louise is a Czech and my ancestors came from Bavaria it was decided that we take a trip to West Germany, Italy, Austria, Czechoslovakia and Switzerland in late summer of 1979. In the winter of 1981 we took a trip to Hawaii. In July of 1982 we took a three-week trip to England, Wales, Scotland and Ireland. This chapter invites you to travel with us as I tell you what we did and what we saw day by day.

WITH retirement out of the way, Schmidt & Company launched under new partners, with settlement of my relationship with Paul and the disposition of our office building, time was available to find a more suitable family dwelling.

Our home in Waldo was comfortable and filled our needs beautifully when we had three small children at home. We needed to be close to school, church, shopping center and public transportation but that no longer applied. There were a lot of steps up and down to enter and exit the house and more steps to the basement and the upstairs. The house was not centrally air conditioned. I contacted real estate brokers and asked them to start looking for a three-bedroom ranch in a nice neighborhood. We found a home in the south part of the city in a subdivision known as Bridlespur. Restrictions have been filed and a homes association implements the requirements. It is a beautiful and comfortable home on a hundred foot lot in a very nice neighborhood. After the purchase of the new house was consummated we sold the old house in Waldo on April 15, 1985 for $50,000.00 cash. We moved into our new home May 5, 1985.

Other than travel in conjunction with business, neither Louise nor I had the time or the money to do much traveling. My retirement from the partnership began January 1, 1978 and from that point forward I no longer conducted audits in the field so it was possible for me to take leave for an extended period. Louise and her older sister attended St. Mary of the Plains Catholic High School in Dodge City, Kansas. After high school Louise returned to Rush County to teach school and her sister Alma entered St. Joseph's Convent. As a member of the order, Alma (Sr. Simeon) taught school for many years in Wichita and when the St. Mary of the Plains College opened in Dodge City, she became a member of that faculty. In 1979 she was college librarian. We received word from her to the effect that some good samaritan made a contribution to the college with the provision that it be held in trust for the benefit of Sisters on their golden jubilee and it was to be used for travel. Sister was awarded the money for travel. That started us thinking about travel plans for ourselves.

I told Louise, "If we are ever going to take a trip, this is a good time to do it. Furthermore, why don't we take a trip to Europe and invite Sister to travel with us?" My ancestors came from Germany, my mother was born in Russia and Louise's father was born in Czechoslovakia so we decided to find some of those places on our trip. All of us got busy on our passports and Sister even obtained her visa. Since I had retained a little of my German language and my brother used his German in Bavaria in World War II, we decided to headquarter in Munich and travel from there to Italy, Austria, Czechoslovakia and Switzerland. We read a few library books on travel in Europe, purchased an European Railroad Pass and booked flights on TWA to New York and then Lufthansa to Munich. With a money belt and a handful of American Express travelers checks (in Deutsch Marks) we took off for Europe on our own. The only hotel reservations we had in Munich was for three nights at the Holiday Inn on Leopold Strasse. We left in early September and planned to end our trip back in Munich three weeks later at the Octoberfest.

We left Kansas City in the morning on TWA, lost some time in Chicago and landed at Kennedy Airport in the evening. We boarded Lufthansa at New York, landed at Cologne in the morning and changed planes for the final leg of the trip to Munich. All flights were smooth and we were especially impressed with the service on Lufthansa. Dinner was served over the Atlantic. We reached our motel Saturday morning

September 8, 1979 and the first order of business was a hot bath and about four hours of sleep. Before purchasing American Express travelers checks, I called Commerce Bank of Kansas City and they suggested purchasing them in German currency. That was a mistake in my opinion. It is much easier to judge values if you compare Lira to Dollars than it is to compare Lira to Deutsch Marks and then Deutsch Marks to the Dollar. As I recall the dollar purchased 1.55 West Germany currency at the time.

Our rooms at the motel compared favorably with Holiday Inns in this country. The plumbing fixtures are different and in our case the electric converter we had did not convert the voltage so we could use some of our electric appliances. We had colored TV in the room and much of the program is American but with German actors and German language. It appeared that nearly everyone in the motel understood English even if they had difficulty speaking the language. Their reservation office was to the side of the check-in counter and when I approached the girls in the office and asked in my best German, "Do you speak English?" One of them remarked, "What do you mean do we speak English? We have to speak English to get a job here." Each of us had at least two bags to carry but our arrangement at Holiday was such that they stored our large bags between trips at no cost. With that privilege we traveled from Munich with a small bag for each of us.

On Sunday morning we attended mass at the Catholic Cathedral and then took a bus tour of the city in the afternoon. The tour guides usually spoke two or more languages. One of the sights I will always remember was a hill or long mound at the edge of the city. The tour guide told us the mound was created with rubble deposited there from the destruction by the American bombing of Munich during World

War II. Photos in the Cathedral provided graphic evidence that the Cathedral took some bombs during the war. All of the building had been restored to its original beauty. Sunday evening we had dinner at the famous Hof Brau House where live music and group singing entertained us. Our dinner cost us fifty marks. The waitress spoke some English and was kind enough to call a cab for us.

Munich, we were told was the third largest city in West Germany with 1,300,000 people. The food at Hof Brau House and at the motel was very good. Munich has a beautiful promenade where no cars are allowed and people stroll as they enjoy the millions of flowers on display. We spent some time there and then purchased tickets for a trip on their subway to the neighboring countryside.

People travel by electric trains in Europe—above ground between cities and underground inside the cities. The subway has elaborate depots complete with shopping centers at street intersections. All European cities are compact with apartment buildings from six to eight stories high. On our trip out of Munich Sunday, we sat near two girls who were speaking English. After introductions we discovered the girls were American exchange students attending the University of Munich.

While strolling through the promenade in Munich, I had a beer across the table from a neatly attired lady and I asked her in my best German, "Do you think the two Germanies will ever unite?" She replied, "No, they will never unite. Our country is too small. In time they will come from the East and from the West and will run over us once more." There are many palaces in Europe and the large one at Munich has perhaps the most beautiful grounds. The palace itself is immense with a thousand rooms. Munich is a beautiful city, full of flowers.

Monday morning we took an Italian Airliner to Rome. There we found an old dirty city. Traffic seemed largely out of control. Keep in mind we stayed at an older hotel near Vatican City and that is the old part of Rome. Food was not as good in Rome as it was in Germany. Prices were probably a little better. Our dinner at a Bavarian restaurant cost about 35,000 Lira. On Tuesday we took tours through the city and through Vatican. The city is full of art objects that the Romans pillaged from countries they invaded during the days of the Great Roman Empire. Some of them from the Greeks and others from Egypt. It was the only city we visited in Europe where we had to fight off pick-pockets. Probably no city in the world preserves its historical past better than Rome. Our tour through the city took us past the fountain where Sinatra introduced the song, "Three Coins in the Fountain." What I said about the City of Rome did not apply to Vatican. It was clean and very much under control. There probably is no greater collection of art objects than is stored in the Vatican. It became very apparent that the Vatican draws tourists by the millions. We had difficulty remaining together with our guide as we went through the main buildings.

Since Vatican City is a separate nation, it has its own currency and its own police force consisting of Swiss Guards. I bought a new commemorative coin just issued with the image of Pope John XXIII and Pope Paul VI stamped on it for safekeeping. We were told that about thirty-five million people visit Vatican each year. The National Geographic Magazine issued December 1985 has a good article and outstanding pictures of the Vatican. Entering the gate to Vatican City with its concrete walls about fifty feet high reminds one of entering a fortress. Incidentally, we did not need a visa to enter any country on our trip except Czechoslovakia. All we needed was our US passport. The church in Vatican City is the largest church in the world and was built over a period of about six centuries. This is the church in which the Pope celebrates mass each Christmas as it is carried live by one of our TV networks. Language was more of a problem in Italy. Very few cab drivers understood either German or English. After a night's rest in our third-rate hotel we tried breakfast at what called itself "American" and like the other meals in Italy it was a poor sample of American food. We took a cab to the Leonardo Da Vinci Airport and returned to Munich about noon. The Israeli check-in counter at this airport was attacked in 1986 by terrorists killing a number of people including some Americans. The area around Rome looks very much like our Western Slope in the United States. Not fertile and rich like the soil around Munich. Probably the most beautiful monument in Rome is the Victor Immanuel marble structure. One appreciates our KCI airport after visiting some of the airports in Europe. As we left Italy, the most thorough search of our baggage took place at customs. We assumed they were looking for drugs. It was the only airport where we saw armed guards as we boarded the plane.

Our first week in Europe was drawing to a close as we returned to Munich from Rome. All of our tours included a lot of walking and we were tired. After a period of relaxation we had a fine dinner at a German restaurant. The coffee was terrible at all of the restaurants. The plum dumpling dessert was great. We slept late at Holiday Inn and then made reservations at the railroad station for our trip to Vienna, Austria. Our passports had to be stamped and while we had a rail pass for all of Europe, we needed seat reservations. Their electric rail coaches have an aisle down one side of the car and compartments on the other side seating about six people. Nonsmoking seats were taken so we had to settle for seats in the smoking section. Louise and Sister were in one compartment and I had a seat in another compartment. On Saturday morning September 15, 1979 we left Munich for Vienna. I had hoped the train would travel through the Alps but it skirted the mountains on the north. We had about a four hour ride through some very scenic country. My companions in the compartment were a couple from Sweden, a man from Brussels, an elderly man from Vienna and a small Japanese boy. The Swede was an engineer attending a convention in Vienna and spoke some English and some German. When we arrived at the station in Vienna, the first problem was to find a rest room and the next most pressing problem was to find hotel accommodations. Unlike the rail station in Munich, the Vienna station had virtually no visible English signs. After locating the ladies rest room marked "Madame," I left our bags with Louise and Sister and headed for the other end of the terminal where a sign read "Motel Information." About ten guys beat me to the information counter. For a time it appeared we might not be able to find a room. The lady in charge must have called about twenty hotels before she found one with rooms available but it was miles from the station. She apologized but I thanked her and we took a cab to the Hotel Ekazent Hietzing. They found a room for Louise and me on one floor and a room for Sister on the next floor for Saturday night with the assurance that Sister's room would be next to ours the following day.

It was a small hotel with a nice restaurant, comfortable beds, bath with tub and shower and the hotel was about one block from a catholic church. We had a good night's rest, attended mass and enjoyed a good breakfast at the hotel. In Europe the hotels arrange for bus tours and add it to your hotel bill. We purchased tickets for a three hour tour of Vienna at the hotel that left the hotel at about two in the afternoon. In Rome, Munich, Vienna and Nuremberg we found remains of what were once walls and motes built around the towns for protection from invading Turks or Romans. As the wall and the mote became obsolete, the ground became a circular street. Like Munich, Vienna was clean and loaded with flowers. Unlike Munich, the terrain is rolling and reminded me of the Ozark Mountain country back home. Vienna is almost as old as Rome and the tour made that clear. There are monuments, parks, palaces, museums, opera houses and streets all over town named after early rulers, artists and composers. One gets the impression that Vienna is a town of music. On the other hand, our bus passed the headquarters of the United Nations in Europe so it also becomes an international meeting place for the East and West. We passed a rather large fountain and the tour guide admitted the fountain was built by the Russians as the allies occupied Austria after World War II. Stalin's statue was removed as opposition to his rule took place in Russia. Our guide told us, "We sometimes tell visitors the monument was built in honor of Jochan Straus." Our tour ended with a tour of another fabulous palace near Vienna. She reminded us that one music student in Vienna became famous later in Bavaria by the name of Adolph Hitler. The group laughed when she said, "American, English, French and Russian troops occupied our country in 1945 and liked it so well they remained for ten years."

We slept late on Monday morning, ate at the hotel and then took a cab to the Czech Embassy to obtain visas for Louise and me. Sister purchased her visa through the mail before leaving the U.S. There were about twenty people trying to get visas and their procedure made little sense to me. It appeared they were taking the applicants in numerical order but apparently that was not followed. I finally reached the window and handed them our applications and our passports. He transferred our passports to another room and waved me aside. I was called finally and after paying the visa fee we were on our way back to the hotel. It must have taken at least an hour to get our visas. I was so glad I knew a little German. Even the numbers were read out in German. Later in the day we booked a tour on the Danube River for 4:00 PM. The boat came into Vienna on a canal and then took the tourists down the Danube to return through the canal to the starting point. We thoroughly enjoyed the peaceful three hour ride down the Danube. Darkness was falling over Vienna and the lights were glistening on the waters of the Danube. Upon our return to the hotel we found a restaurant nearby by the name of "England" where we had a nice dinner.

After breakfast at the hotel Tuesday morning September 18, we booked another tour of Vienna that took us to St. Stephens Cathedral, the Spanish Riding Academy and a fine arts museum. St. Stephens, like so many churches in Europe is about a thousand years old and very ornate. The ceiling must be a hundred feet high and the steeple 450 feet high. Photos taken after World War II indicate that much of the church was destroyed by allied bombing. In fact we were told that thirty percent of Vienna was in ruins after the war. Since the catholic monarchs of the day built the churches, many of them served as mausoleums for the royal family. Like Munich, Vienna was completely restored after the war.

The Spanish Riding Academy was also property of the monarchs originally as their private riding stables. Later it became a tourist attraction as the famous training school for the performing white Arabian horses on tour all over the world. Like so many of the historic places in Europe, the buildings were not designed to accommodate millions of visitors. There were no elevators in the large building and the concrete stairways were narrow and long. We were able to watch as many as six horses being trained in the main arena at one time. The place was clean and both the horses and the riders were in colorful attire.

Judging by the precautions taken at the fine arts building, the contents are probably irreplaceable. We were not allowed to carry a coat or a camera into the rooms where the paintings were on display. We were allowed to wear a coat but not to carry the same. We returned to the center of town to a nice promenade but they did not have as many flower pots as we saw in Munich. After dinner at the hotel, we started putting our things together for our trip behind the iron curtain.

There is an interesting side-light to our stay at Hotel Ekazent Hietzing. An elderly white haired gentleman by the name of Roman Maderbacher was at the desk when we checked out Wednesday morning. The hotel had posted our four night's stay in pen and ink in three columns and accumulated the total in a fourth column. The total bill came to 3,729 shillings for the room and 750 shillings for the tours. I paid Roman with a 500 DM American Express check and 929 shillings. At the time 500 DM were worth 3,550 shillings. When we got back to Kansas City I had a letter from Mr. Maderbacher. The letter was dated September 21, 1979 and written in German long-hand. I can read a little German when printed with English letters but only my mother could read German in long-hand. He was kind enough to insert some English words:

> *Herr Schmidt—it is completely my mistake and fould-ven you check your account, your vill see yourself vats happen. Your bill should be A. S. 4439. I charged you 3729 shillings. So I made a mistake of 710 shillings. (Approx U.S. $60). Please send me the difference ($60) in a letter or*

> *check, otherwise I have to pay myself to the hotel. You speak verry good German, therefore I will tell you. Herlichen Dank in advance.*
>
> *Roman Maderbacher, Receptionist*

On October 2, 1979 I replied:

> *Attached are photo copies of both sides of the bill we paid. As I recall I paid you with a 500 DM American Express check and 929 shillings. It looks like you found your mistake and added it to the amount on the back side of the bill. If anything we overpaid 40 shillings.*
>
> *We enjoyed our stay at your hotel and wish you the best of everything.*

Sometime later I received another message from Roman in which he mailed what looked like the original ledger sheet and with a red (x) he noted "X-here, I made the mistake in addition. Please check your bill again."

When I checked the bill I saw his mistake in accumulating the daily charges in the last column of his ledger sheet. He made an error of 710 shillings.

On October 23, 1979 I wrote Roman Maderbacher:

> *Dear Mr. Maderbacher:*
>
> *There is a Chinese saying, 'He who*

makes no mistakes, does nothing. He who makes too many mistakes gets fired.'

Mr. Maderbacher, what is right is right and the Atlantic Ocean does not change that fact. According to the Commerce Bank in Kansas City the U.S. dollar converts to thirteen shillings this morning. Accordingly, my check for $55.00 is enclosed in payment of your inadvertent error of 710 shillings. I apologize for not checking the bill while we were at the hotel. On the other hand, by the time we left Vienna, I was thoroughly confused as to the value of Deutsch Marks, Italian Lira and Austrian Shillings. It became more confusing when I traded Austrian Shillings for Czech Korunas and Deutsch Marks for Swiss Francs.

Be that as it may, we enjoyed our trip to Europe and would not hesitate to return to Vienna to stay at your fine hotel.

Among the Christmas cards we received in December 1979 was one from Roman Maderbacher in Vienna.

After checking out of our Vienna hotel, we took a cab to what we were told was the rail connection to Brno, Czechoslovakia, but I soon learned we were at the wrong station. A little later we arrived at the correct departure point for our trip behind the iron curtain. Our rail pass was good to the Austrian border but we had to purchase railroad tickets for our trip through Czechoslovakia. The passenger train had only about three cars. We soon arrived at the barbed wire gate patrolled by armed soldiers and six armed people in uniforms approached us in our cabin. After asking for our visas, they wanted to know where we were staying. We told them we planned to stay at a hotel but we had no reservations. They queried us at some length and we kept repeating that we were visitors and had no hotel reservations or relatives with whom we planned to stay. A girl among the group asked us to purchase 320 Korunas with foreign currency. We used 528 Austrian shillings in payment and the girl posted the amount in each of our visas. We were asked to spend the currency before we left otherwise we would have to return the balance. We suspected that at least one in the group understood English but they were careful not to divulge it.

Brno, Czechoslovakia was selected for our first stop because it was the nearest rail junction to the place where Louise and Sister's father was born and where their grandfather lived. The trip to Brno was pleasant through rolling country but we noticed very few farm dwellings. Large groups of what appeared as school children were picking potatoes. We saw some irrigation, corn, sugar beets and the rolling hills were covered with grape vines. As the train passed a railroad station on the way, we noticed a group of teen age boys with packs on their backs waiting to leave. We assumed they were being drafted for the Russian military. Upon arrival at Brno our first contact was a hotel but we were told we would have to return to CEDOK the state-owned station to register for a room. When I arrived back at CEDOK the scramble for rooms repeated itself. We were told an international convention was in progress and very few rooms were available. We were fortunate in that one of the girls in the office spoke some London English and some German. She finally found rooms for us in a university dorm at the edge of town that had been converted to a hotel. When I wanted to pay for two night's lodging I ran into trouble. The cashier would not accept Korunas in payment and did not trust my American Express travelers checks issued in German Marks. I finally noticed a VISA sticker in her window and offered my Visa card issued by Commerce Bank. She took my card to a back room and finally accepted it as payment for the rooms. During this time I shuttled from window to window and each time a girl had to put an ink stamp on something. The English speaking girl commented, "Oh we love our red stamp here." What a difference between the city of Brno and cities like Munich or Vienna. The entire physical plant of Brno seemed in disrepair and exhausted. We took a cab to the University for our rooms and as the driver followed an old delapidated, smoking truck he commented in Czech language, not realizing that Louise and Sister understood Czech, "That is what we get from the Ruskies, noise, dirt and stink." Our rooms at the University were small, as most rooms are in school dormitories, but we had two rooms with single beds and a common shower. The beds were clean and rather comfortable. Weeds had grown up around the building but the interior of the building was clean. On our way to the dorm we noticed people lining up at abandoned buildings to purchase vegetables being offered by individual farmers. It seemed like school age children were supervised by scouts and not by parents. Only real small children were with their mother. Wednesday night we ate at a nearby tavern serving food. The customers drank more than they ate. We had pork and sauer kraut and it was fair. The coffee was terrible. The cost in U.S. dollars was four or five dollars per meal. A little store near the dorm was not well stocked and had few labels we recognized.

Thursday morning we made it to the first floor of the dorm where we found a small cafeteria. A rather large number of guests appeared to be engineers from other countries. The only food served was what they called, "Am & Eg." Most of us had a coke instead of coffee. After breakfast we took a cab back to CEDOK and the first order of business was the purchase of railroad tickets to Prague on Friday. Someone told us we would have to purchase the tickets at the railroad station about ten blocks away. I left Louise and Sister at CEDOK and walked to the railroad station only to be told we were foreigners and would have to purchase the ticket at CEDOK. To purchase a ticket, one applies at one window, stands in line at another window to pay for the tickets and stands in line at the third window waiting for the tickets. The hassle took the better part of two hours. With that chore accomplished we asked the attendants to find an English or German-speaking cab driver to take us to the area where the Pivonka's used to live (Yamaritza). They were not certain that this could be obtained but finally a driver offered to take us but he knew no English and very little German. We could have rented a car but our time was so limited and we probably would have encountered difficulty finding the place.

The first part of our trip out of Brno was on a divided highway similar to our Interstates in the US. Brno is a city of about 350,000 people and the larger cities are connected with divided highways. The last part of our journey was through rolling country, a lot of curves and narrow dusty roads. There were clusters of what appeared like Pine trees and then next to these small forests was beautiful farm land. Since there are so few farm buildings, we assumed the people must be hauled out from the small towns to work the farms and then returned to the towns at the end of the day. A construction crew was working along the road supervised by an armed guard. Our first stop was at Yamaritza but the Pivonkas lived in a smaller hamlet about six miles from town. Unfortunately, our driver had no idea where the hamlet was located. Sister finally wrote the name of the hamlet on a piece of paper and handed it to the driver. The driver jumped from his cab and some local people directed him to the little hamlet. In the mean time, the three of us visited the church attended by the Pivonkas. Like so many churches in Europe, it was ornate and beautiful with marble everywhere. The ceiling was painted almost like the Cistine Chapel in the Vatican. The pews on the other hand were wooden and very uncomfortable. The little hamlet was virtually deserted with the exception of two small children and an elderly man returning empty beer bottles to the store. There was a small chapel in the center of the hamlet, just as the Pivonkas had reported. Louise and Sister visited with the old gentleman in their limited Czech language and he told them he remembered the Pivonkas well. He pointed to a house on the corner of the block as the location of their house but commented the original house had been replaced. He told the girls the Pivonkas had migrated to the United States where they became wealthy. This brought a chuckle from Louise and Sister. The cab driver and I had agreed on cab fare but when we returned he offered to settle for less if I paid him in foreign currency. Instead of returning to the dorm upon our arrival we killed a little time in downtown Brno and had dinner at a local hotel coffee shop. It was a warm day but the waiters were in black suits and bow ties. The food was not all that fancy.

Friday morning found us on the train to Prague where we planned to spend two days. We were blessed with good weather on our trip so far but on that morning it changed to rain and colder weather. If anything, the terrain between Brno and Prague is more rolling than the area from Vienna to Brno. Prague, the capital of Czechoslovakia has over a million people and like most cities in Europe has an elaborate subway system. In Prague, however, the central railway station for the long-distance trains also serves as the central station for the subway system. It was easy to get confused. Our first need was hotel accommodations and as usual we found it difficult. The station had a large sign "Information" posted in the window but the character behind the window knew not one word of English. He did point to a room across the way. I made my way to that area and luckily a girl in that room spoke some English. She asked me what part of the United States I was from and I told her. I asked her if she had been to the United States and she replied, "No, but I sure wish I could." Once more the hotel reservations were handled by CEDOK. One of the girls called an office that handled class A hotels but they had no rooms available. She did find a room at another office for class B hotels. Whereupon, we took a cab to CEDOK for class B hotels.

We found a girl at that office who knew some English. Unfortunately, the only room available was a large room with three beds and bath. It was getting late so we took the room. I cashed a hundred Mark travelers check for Korunas at this place and it seemed like I received a lot of Korunas. While this was taking place an elderly lady approached Louise and was offering her twenty to one for American currency. The official exchange was about eight to one. She wanted us to stay in her home but we declined. After quite a long cab ride over the river we arrived at our hotel. The room was on the fourth floor and with three single beds. The bath room consisted of a shower so we had to use public rest room facilities around the corner. The elevator worked only part of the time so we had to walk stairs. It was undoubtedly a class B hotel. We were fortunate in that the elderly hotel clerk spoke good German. We took the cab driver's advice as to an eating place and had dinner at a tavern not far from the hotel. As we came back from dinner, the hotel clerk called me in German, "Herr Schmidt, I have a telephone call for you." He handed me the telephone and I determined the caller was the girl who had made change for me at CEDOK. I handed the phone back to the clerk and he talked to the girl in Czech and after doing a little computation on a scratch pad he explained the girl gave me the correct change for American currency but the currency I gave her was German currency so she gave me too many Korunas. We told her if she could get to the hotel I would refund her the difference. She took the tram to our hotel at about ten o'clock in the evening, came to the fourth floor, made out a new ticket, tore up the old ticket and I refunded her in American money. She went away happy. Besides being such an undesirable room, we had another problem, the reservation was for only one night. I asked Louise and Sister whether they would like to stay the two days as planned and offered to go in search for other accommodations. Both had reached the conclusion, they had all of Czechoslovakia they cared to see.

Our reservations at the Holiday in Munich was for Sunday night so we elected to take the train to Nuremburg Saturday and then to Munich on Sunday. It was still raining when we came to the hotel lobby at five in the morning and asked the clerk to call a cab for us. He dialed and dialed and got no answer. He told me in German, "The night crew must have gone home and the day crew has not arrived." The old gentleman even stood in front of the hotel hoping to flag a cab for us to no avail. At about six o'clock he finally got some response to his telephone call. We arrived at the railroad station in time and were able to purchase our ticket to the border of West Germany where our pass was effective. The eating place at the station did not look appetizing so we purchased some canned fruit juices and boarded the train for Germany. We soon learned the communists take great precautions with people arriving through the iron curtain but they take even more precautions to make certain no one leaves through the iron curtain. Our train was halted at the border for over two hours.

Outside the train was the iron fence, armed soldiers were walking back and forth guarding it and once more six people boarded the train to see our visas. We had forty Korunas in bank notes to return. I had a lot of change but they were not interested in the return of the coins. While the train was standing, people came into our cabin with flashlights and using the window as a step ladder removed the ceiling of the car to make certain no one was hidden above the ceiling in the car. The lid of a railroad tank car was removed on a freight train next to our train and people were searching the interior of the tank car for escapes. Others were crawling under the rail cars searching for escapes.

The three of us gave a sigh of relief when our train started rolling across West Germany toward Nuerenberg. The railroad tracks were much smoother, the rural areas were dotted with farmsteads brightly painted and we were in a different world. We were tired and hungry by the time we reached our destination but happy to be out of Czechoslovakia. At the railroad station in Nuerenberg I asked a young lady to find us a double and single room with bath in a nice hotel and within minutes she found rooms within walking distance of the station. We walked into rooms nicer than the ones we occupied at our headquarters in Munich. The town was buzzing with activity because the Octoberfest was in full swing in Munich and Nuremberg has its own Octoberfest at the same time. There were bands playing and concession stands were scattered over the promenade. At the suggestion of the hotel manager, we had a fine dinner at the Holy Ghost restaurant dating back to the thirteenth century. Next morning we attended mass at St. Clara Catholic church almost next door to the hotel and started packing for our return to headquarters in Munich. At first we were not so sure St. Clara was a catholic church. The building at one time was home for an order of nuns, we learned. The mass we attended was celebrated in what was the chapel. The priest arrived in a business suit with a brief case under his arm. He shook hands with me and an elderly lady was getting the chapel ready for mass. I finally asked her whether this was a Roman Catholic church and she assured me it was. A later mass was celebrated in the main part of the church. The train ride to Munich was relaxing and we arrived at Holiday Inn on Sunday evening September 23, 1979. Due to the Octoberfest, our reservation at the hotel was for only one night so we planned to take the train to Switzerland on Monday.

Our train took us through a corner of the little country of Lichtenstein before we reached Zurich, Switzerland. Checking our passports in this case was mostly a formality. Our rail companions in the cabin was a couple from Minnesota. A nice couple on their honeymoon. According to them, they were neighbor kids playing in the same sand pile, married and went their separate ways but after the loss of their spouses found each other again. They came from Paris and reported France was not as nice as Germany or Switzerland. On our way to Zurich, we saw some beautiful meadows, orchards and vine covered hills. Sheep and cows were grazing under fruit trees in the meadows. Upon arrival at the Bonhuf (railroad station) I walked to the cage in the center of the station and asked a young lady in German, "Do you speak English?" She responded, "Why, you speak good German." I told her we were interested in hotel reservations and she directed me to follow a red line painted on the floor of the station to the hotel reservation in another part of the building. After converting some German currency to Swiss Francs, I made my way along the red line to the hotel reservation office. I asked for two nice rooms with bath and she apologized for having no rooms in a four star hotel. The best she could find were two rooms at the Ambassador Hotel which was a three star hotel and not close to the station. I assured her that was no problem because we would hail a cab to the hotel. She wrote the name of the hotel and address on a piece of paper for me to hand to the cab driver. When we arrived at the Ambassador, we discovered the nicest rooms we saw in all of our travels. From the double front doors for safety, to the beautiful glass doors leading to the balconies, to the mini-bars in each room, they were strictly first class rooms. The same was true in the coffee shop where we had dinner. The quality of service, the silverware and the quality of the food were excellent. This was the only place where the hotel maintains a bulletin board at its front counter providing the guests with the market value of the currencies of about ten or twelve countries in Swiss Francs. We looked at some tour bulletins and finally selected one to the black forest of Germany.

We boarded our tour bus at the opera house grounds about a block from our hotel. There were about thirty passengers on the bus representing twelve different nations. Our route was almost straight north from Zurich and followed the Rhine river. Our passports were checked as we entered Germany and we stopped for lunch in the heart of the black forest. The Rhine flows to the north and several dams have been constructed to provide large recreational areas for fishing, boating and water skiing. We visited one of the largest churches in Europe, with the exception of the Vatican, in a little town along the Rhine. It was constructed in about 800 AD, was operated by the Benedictines for a time, at one time it was used as a factory and then the Jesuit order gained control of it and now it is being used as a boys school. I have no idea why the forest is called the Black Forest because the pine trees look just like the lodge pole pine or the ponderosa pine of Colorado. We were fortunate in that our tour guide did quite well with the English language. In addition to seeing some beautiful country dotted with quaint little villages every ten miles or so, I was able to learn more about the country. It is a very small country of about six million people. When she told us they had a standing army of a half million people, my question to her was, "How can such a small country organize an army of that size?" She explained that every man to the age of fifty is obligated to take military training, store arms in his home and be available for military duty within forty-eight hours. Even the disabled are required to serve in a non-military phase in the event of an invasion. Perhaps that is why Switzerland has remained at peace and neutral over the past two hundred years. We learned that Switzerland has no fossil fuel such as coal, gas or oil and must rely on water and nuclear energy for its electricity. Keep in mind this was in 1979 and, at that point in time, 70% of its electricity was provided by hydro plants and thirty percent was supplied by nuclear plants. I expected to find nothing but mountains and rocks but the part of Switzerland that we saw was green rolling terrain loaded with forests, orchards, vineyards and meadows. The city of Zurich has about 400,000 people we were told. Our trip along the Rhine passed a number of castles but we did not stop to visit them. We were told that use is being made of some for conventions of various types. This was a very enjoyable trip and we arrived safely at our rooms at the end of the day.

Wednesday, September 26, we did a little shopping in downtown Zurich before we took a cab back to the rail station for our return to Munich. The prices in Switzerland were about the same as they were in Germany. Our trip back to Munich was enjoyable and after we arrived at the Holiday Inn we decided to visit the Octoberfest. The five hour trip was shortened by the fact that a couple from San Francisco shared our compartment on the train. We wondered how they hoped to find hotel accommodations without a reservation during the Octoberfest.

After checking in at headquarters we took a cab to the Octoberfest. We were rather tired but our plans called for a tour south of Munich to the little villages on the side of the Alps for Thursday and we did not want to leave without at least getting a glimpse of it. We were told the affair has been held for about two hundred years on eighty acres donated to the city by the Hapsburg family. Munich has about seven beer breweries and they put on the show. We probably picked the most crowded night because it had been raining for days but the sun was out on Wednesday. There must have been about one hundred thousand people jammed into the beer tents and concessions erected on the grounds. We stopped at the Lowen Brau tent but were unable to get inside. There were some tables in front of the tent with overhead electric heaters and we found room at one of those. Lowen Brau was serving broiled chicken, bread rolls and beer. After serving the chicken, the beer and bread rolls kept coming. The standard beer mug was one liter. The half-chicken was broiled in the tent on a vertical electric rotisserie about six feet high and twenty feet long. Louise and Sister decided to remain in front of the tent but I squeezed into the tent to where the action was. The enclosure was hundreds of feet square with a raised band stand in the middle. Tables and chairs covered the ground. A German band of about twenty members entertained. One of the band members told me the tent had a seating capacity of 5,000. As the band played the throng would stand, hold their beer mugs in the air, sing along with the band and sway back and forth like a Kansas wheat field. It was crowded and it was noisy. But the Germans were having a good time and while I saw security guards in the wings they seemed to have no trouble controlling the mob. This party draws millions of visitors who drink millions of gallons of beer over a two-week period each year. One night was enough for us as we returned to our headquarters. We were told that erecting the tents and concession stands for this celebration takes about three months and about the same time is required to remove and store them for the next year.

On Thursday morning September 27 we had breakfast at our motel and boarded a larger bus for our trip toward the Alps. The capacity of the bus was about sixty people and it appeared all seats were occupied. While we loved the scenery in Switzerland, this tour was probably even more beautiful. The hills and valleys were covered with green, the little towns brightly painted and covered with flowers. We were told the highest peak in that area is a little over 9,000 feet. In other words, not as high as the rockies in the United States. We were near the Austrian border but remained in West Germany for the entire tour. Our tour guide spoke English, Spanish and German. We had lunch at a little town by the name of Garmish where the Olympics were held several years ago. One of the highlights of our trip was visiting the Lindau Castle built by King Ludwig II. Our Gastriech neighbors in Kansas City referred to him as Crazy Ludwig. He wasn't crazy after all. West Germany reaps millions of dollars from tourists visiting his castles. It was one of the smallest castles we visited and one of the most ornate. We were told he was a friend and admirer of Louis XIV of France and copied much of the French architecture. The interior was covered with gold and silver. The guide told us Ludwig II was also a close friend of Richard Wagner and often invited him to the castle to play on the musical instrument we saw in the castle. It appeared like a combination organ and piano. We also visited the little town of Oberammergau where the famous Passion Play originated and is being performed to this day. A human plague hit that part of the world in the thirteenth and fourteenth century referred to as the Black Plague. People were dying in great numbers and there seemed no cure for the disease. People flocked to their churches and to the hills praying for Divine assistance. It is said a promise was made that if the scourge of the plague could be lifted from the people, the passion and crucifixion of Jesus Christ would be reenacted every ten years in this location. We were told that plans were underway to perform the passion play in 1980. When we visited St. Stephens Church in Vienna, we saw a large gold and silver monument in front of the church erected in appreciation of the end of the black plague. On our way home from Oberammergau we visited a Benedictine Monastery that claimed to have a secret formula for wine they were offering to sell the tourists. We reached our motel tired and hungry. After some roast duck at an old Czech Restaurant we returned to our headquarters and made plans for our return to the United States.

Friday September 28 we returned to Kansas City by the same route used for our trip to Europe. A problem developed in our connection between Lufthansa and TWA at Kennedy in New York. In the first place, Lufthansa checked our bags through to Kansas City and I assumed they would not have to pass through customs. In the second place, the scheduling of our TWA flight home allowed only one hour to make the transfer in New York and that was not enough time. The airport was jammed, there were no porters available to help with the bags and long lines of travelers were awaiting inspection by customs. The cab that was to wisk us to the TWA terminal had to stop short of the terminal because of a traffic jam, due to ambulances trying to remove someone who had collapsed in the terminal. There was a long line at the check-in counter at TWA but a TWA employee helped us bypass the line to get to our plane. If the plane had been on schedule we would have missed our flight home. There is something to be said for non-conducted tours. This was especially true of our trip behind the iron curtain. I am certain we saw and experienced things that we would not have experienced with a conducted tour. It is a more leisurely trip than one in which the entire tour is predetermined. The three of us never regretted our non-conducted tour of Europe in the fall of 1979.

As stated before, my older brothers and sisters moved to Chicago during the depression years. My oldest brother, Ben, passed away May 31, 1966, but the next oldest brother and fisherman, Henry, is retired and lives in Villa Park, Illinois, a suburb of Chicago. During the hot summer of 1980 Louise and I called Henry and Rose to see if they would be interested in a tour around the Great Lakes, hoping to find some good fishing places. They accepted, so we started packing. Our first day out of Chicago took us by Notre Dame University and on to the Wolverine Motel at Ann Arbor, Michigan. Henry & Rose had not visited Niagara Falls so that was our next stop. The Canadian side of the Falls is so beautiful. We spent the better part of the afternoon there and took a bus tour around the park. So much water falling two hundred feet is breathtaking. After visiting the falls we drove around Lake Ontario and found a nice motel east of Lake Huron. We did not drive into Toronto, but the highway took us around the outskirts of the city. The temperature was in the seventies and eighties while the temperature in Kansas City hovered around the 100 degree mark. It was a beautiful place to spend a week or two. On our fourth day out we reached WaWa, Canada, on the north shore of Lake Superior. Sunday morning we attended mass at a little catholic church in WaWa. We noticed that one of the masses was celebrated in French. Our entire trip around the lakes took us through a panorama of rolling terrain, forests and blue lakes. A most beautiful drive. As we started down the western shore of Lake Superior, Henry and I looked for a nice fishing spot. We did find some real nice fishing lakes, but unfortunately no fish were caught. On our way back to Chicago we stopped at Menasha, Wisconsin, where some of our cousins are in business. It was a short trip but a most enjoyable one for all.

After the books were closed for the calendar year 1980, I suggested to Louise that we consider a trip to Hawaii. It is, after all , our fiftieth state and we had never been there. This time we opted for a Four-Island Hawaii Silver Circle tour conducted by the Grand Circle Travel Agency. All travel arrangements were made by the agency from the United Air Line tickets, the Hawaiian Airline tickets, the hotel reservations, the bus tours, the meals and the entertainment. Language was no barrier on this tour. Twenty-eight of us arrived Monday evening, February 16, 1981 at the Hyatt Hotel near the Los Angeles Airport for a briefing, introduction to the tour guide, getting acquainted with fellow travelers and enjoying dinner together. Next morning we took a Boeing 747 to Honolulu where we were greeted with the traditional flower lei and a kiss of aloha. There was a slight change in our travel plans, so we had to take the Hawaiian Airline to our first destination, Hilo on the east shores of the large island, Hawaii. An elderly couple from New York was seated near us on our flight to Honolulu and as the 747 made its approach for landing and the high-rise buildings came into view, the man commented, "Looks just like New York." The weather was just perfect and flowers were everywhere. Our flight to Hilo was smooth and before long we checked in at Billy's Hilo Bay Hotel. According to Billy, it is the only family-owned hotel on the islands and Billy told us he and his family built it from scratch. The hotel had 150 rooms and our room was clean and comfortable. Billy and his family entertained us at dinner the following evening with some Hawaiian music and song. There are tourist shops everywhere, the stores have no doors and the same was true of our dining room. The soft ocean breeze flows through all structures including churches. Louise and I bought some bright colored apparel for our tour of the four islands.

On Wednesday we took a bus tour to Hawaii Volcanoes National Park on top of 13,000 foot Mona Loa mountain of volcanic ash. On the way we visited Liliuokalani Park with its sparkling Rainbow Falls and an orchid nursery. After walking around the rim of Gigantic Kilauea Crater we had lunch at the Volcano House Restaurant in the park where we met a couple from Estes Park, Colorado. We noticed a rather large federal observatory installation on the top of the mountain that monitors the activity in the volcano. We also visited a museum of volcano exhibits. After our return to Billy's hotel at about three, we had a nice nap and then attended dinner at the hotel where Billy and his family entertained us.

On Thursday morning we left Billy's for a tour of the north half of big island on our way to Kona on the western shore of the island. We packed our bags and placed them outside our hotel door where the bus service took care of the bags. The north half of the island also has a 13,000 foot Mona Kia mountain. As we followed the northeast shoreline we passed a number of sugar cane plantations. This part of the island receives about 100 inches of rain each year. We passed the Parker Ranch, the second-largest cattle ranch under the American flag. It was very obvious that the western slope of the island is the desert side. Before checking in at the Kona Hilton we took a short tour south to the City of Refuge at Honaunau. This sanctuary was built in the twelfth century as an asylum for fleeing warriors and taboo violators. The Kona Hilton is one of the finest hotels we have been in. Kona is on the leeward side of the island with beautiful beaches.

On Friday morning our tour group reserved space on a good-sized boat named Captain Cook for a Kona Coast Cruise. It was a very relaxing ride down the west coast of the island and stopped for pictures and swimming at Captain Cook's monument. The water on this side of the island is blue as the sky and clear to depths of fifty or one hundred feet. From the air the ocean on the east side of the island appeared green. The Kona Bay area is becoming a popular resort area for many Americans. Louise and I had a light lunch at the Hilton and strolled through the area for some photos in the afternoon.

On Saturday February 21, 1981, we took the plane to the island of Kauai with a stop on the island of Maui. Maui is the second largest of the islands. We took a tour of the island starting with the 1,200 foot Iao Needle, a fern-covered rock formation in the Iao Valley, then we took a scenic drive to Lahaina. This charming settlement was once the whaling capital of the world. An old restored whaling ship was anchored nearby for us to enter and examine. We also sat under the largest banyan tree in the United States. Like the west coast of Hawaii, the west coast of Maui has beautiful beaches and is a very beautiful resort area. All of America's first class motels are there. Louise and I indulged in a little ice cream at Lahaina and strolled through the small settlement. Later in the afternoon we returned to the airstrip and finished our flight to the island of Kauai.

Our Kauai Resort Hotel was strictly first class located on a beautiful beach lined with cocoanut trees. We enjoyed the balmy breeze and the pacific white caps from our balcony. On Sunday morning we took a cab to church, a rather large metal building without doors. The mass was celebrated in English accompanied by a guitar ensemble. In the afternoon we took a barge ride up the Wailau River to the haunting splendor of the

Fern Grotto, a cave similar to some in the Ozarks of Missouri. Sunday evening we and perhaps two hundred other tourists were treated to a luau. This was provided on the grounds of our hotel complete with hospitality bar and a buffet dinner accompanied by Hawaiian dancers and music. It was quite long and loud but the food was good and we enjoyed it. Monday morning we slept late and then took a bus tour to Waimea Canyon referred to as the Grand Canyon of the Pacific. The island of Kauai receives about 400 inches of rain in the higher elevations each year and that probably accounts for the erosion over the years to create this spectacular canyon. Monday evening we flew back to Honolulu where we checked in at the magnificent Hilton Hawaiian Village located on Waikiki Beach. This would be our headquarters for the remainder of our stay in Hawaii. We had a beautiful room on the sixteenth floor, complete with parlor and balcony. Our Monday evening meal was a Chinese Banquet Dinner. Chop sticks were provided but most of us soon switched to other silverware.

The state of Hawaii has about one million people and about half of them are on the island of Oahu where the capital of Honolulu is located. On Tuesday, February 24, we took a tour of Honolulu including the state capital and the Bishop Museum, but the National Memorial of the Pacific impressed me the most. The memorial is located on a flat topped hill a short distance from downtown Honolulu. The hill is an extinct volcano crater estimated to be 75,000 years old. In 1949 the 112 acres on the floor of the crater were dedicated as a National Memorial of the Pacific. More than 18,000 war casualties are buried here including veterans of the Spanish American War through the crisis in Viet Nam. Ernie Pyle, the well-known World War II correspondent is among them. A monument in the memorial called "Garden of the Missing" has this poignant inscription, "The Solemn Pride That Must Be Yours—To Have

Laid So Costly A Sacrifice On The Altar Of Freedom." Engraved on its marble wall are the names of thousands who lost their lives in World War II.

Tuesday evening we were treated to dinner on a Pagoda Floating Restaurant. The dinner consisted of Japanese cuisine and was good. On Wednesday we took a tour to the north end of the island of Oahu to see some popular surfing areas and as we rounded the tip and started down the east coast we stopped at the Polynesian Cultural Center maintained as part of Brigham Young University. We visited displays of about eight different cultures of the Pacific and were treated to a good dinner on campus followed by a fine performance of Polynesian dancers and entertainers. The bus brought us back to our hotel at about ten o'clock. Thursday, February 26, was a day for relaxing. We slept late and strolled on the beach of Waikiki. That evening we were treated to a dinner playhouse on the Hilton grounds. One of Honolulu's most entertaining shows was performing, "Jim Nabors' Hawaiian Review." Both the dinner and the show were super.

On Friday we took a tour up the east coast of Oahu, very rugged but scenic. Our final destination was what was termed, "America's Finest Display of Marine Life" at Sea Life Park. It was a windy day and we noticed a number of school children at the park. We had lunch at the park and returned to the hotel in the afternoon. The Hilton had about three towers and was building a fourth while we were there. For dinner we tried a dining room in one of the other towers facing the beach. We had a fine meal of prime rib and lamb. We were pleased to find the cost was about the same a similar meal in a first class eating place in Kansas City would cost. Saturday was another leisurely day to enjoy to the fullest. We did some strolling through the shopping centers and on the beach. Saturday evening we were treated to a dinner on Hilton's own Rainbow Catamaran. Boat motors were started to push the boat out to sea for a short distance and then motors were silenced and sails raised. We sailed along Waikiki beach for about two hours enjoying complimentary drinks and a fine dinner accompanied by two Hawaiian Guitar players. It was a nice way to end the day. Our two-week tour was drawing to a close.

We attended Mass in the hotel on Sunday morning at nine o'clock. What was a chapel in our tower was being converted to a restaurant, so temporary facilities were used for the mass. At ten thirty a protestant service was conducted in the same room. A new chapel for all denominations is to be part of the fourth tower. The attendance at mass was about as large as attendance at our church in Kansas City. The protestant service was not as well attended. Our last tour was a boat tour on Sunday to Pearl Harbor. It was a fully narrated tour passing by Battleship row, the sunken USS Arizona and USS Utah vividly recalling the tragic December 7, 1941, attack on Pearl Harbor. There were many Japanese on the tour. Millions of people come to see Pearl Harbor each year. We passed, but did not enter, the USS Arizona Memorial. The memorial was dedicated in a solemn ceremony on Memorial Day 1962. It spans the sunken hull of the Arizona where the bodies of more than 1,100 men are still entombed. Within the Memorial is a museum, a shrine where the names of the Arizona crewmen who died are inscribed on a marble wall. The assembly area is large enough to accommodate two hundred people for memorial services. The terrible destruction wrought on Pearl Harbor on december 7, 1941 came from a 33 ship Japanese task force which had crept undetected within 200 miles north of Oahu. The attack of 353 planes from six aircraft carriers began at five minutes of 8:00 o'clock Sunday morning and by one o'clock the devastation was complete and the task force was leaving the area for Japan. The attack brought a Declaration of War from the United States against Germany and Japan.

Upon our return from Pearl Harbor we had our farewell dinner at the Hilton and started packing for the return home. Many of our tour group were returning with boxes of pineapple. Bill Bertrand, our bearded tour guide, did an excellent job of taking care of our every need. By four o'clock Monday evening we landed back at Los Angeles. Instead of flying to Kansas City that evening, we enjoyed dinner with some tour friends and spent the night at the Hyatt Hotel near the airport. Both Louise and I agreed this trip was much easier on us than our trip to Europe. It was nice to be able to visit with our tour buddies who hailed from New York to Montana. A leisurely, beautiful trip.

Since we enjoyed our Grand Circle Travel Agency conducted tour to Hawaii in the winter of 1981, Louise and I started looking at other tours sponsored by the same travel agency. We finally selected a three-week summer tour of England, Scotland, Wales and Ireland, to leave Kansas City July 23, 1982. We left Kansas City about noon Wednesday on TWA to Kennedy in New York by way of Detroit and then flew across the Atlantic Wednesday night on an India Airline Boeing 747. Dinner was served over the Atlantic and the food was good. Unlike our nice 747 to Hawaii, this plane was poorly maintained, not very clean and loaded almost entirely by Indian people from Bombay, India. We met Irene, our tour guide, Thursday morning at London Heathrow Airport and were taken with our bags to the Kensington Palace Hotel in downtown London where we hoped to stay for three night. The hotel was an old structure but apparently refurbished on the inside. Our room with twin beds was very nice. We rested most of Thursday. Friday morning we visited London Towers with its color and pomp. Next we visited Westminster Abbey and St. Paul's Church. Most of the churches, and there are many churches in England, were built long before Henry VIII. Many Roman catholics are buried inside these churches. After touring a good part of downtown London, we were treated to an excellent get-acquainted dinner aboard a boat anchored in the Thames River. On Saturday we visited Windsor Castle about thirty miles southwest of London. We were told the flag atop Windsor Castle meant the Queen was in residence. The castle, like so many in Europe, sits on a hill and what appeared to have been a mote surrounds the hill. We saw the change of the guards at Windsor Castle and it was a beautiful day. Saturday evening we attended a play in downtown London. Since our tour was to leave London rather early Sunday morning, we were interested in finding a church near the hotel and an early Sunday morning mass. The Kensington Hotel receptionist directed us to a roman catholic church within ten minutes walking distance from the hotel. We attended mass at 6:30 in this large Carmelite Church and the celebrant was definitely Irish. It must be a rather large parish in view of the number of masses on Sunday. After mass we had breakfast at the hotel where we met a very interesting couple in the coffee shop on vacation from New Castle, England. One of their children had been to Atlanta, Georgia. While we were wearing sweaters nearly every day, the sun was out as we boarded our bus at eight o'clock. Our next stop was Bath, England. Enroute we stopped at Salisbury, where we saw another beautiful church. We stayed at the Francis Hotel and in a very comfortable room. On the way to Bath we visited the Stonehenge, a unique structure of unknown origin. Stonehenge consists of a triply concentric group of stones arranged nearly in a circle and within the circumference of a deep trench, the diameter of which is 333 feet. Numerous theories have been advanced over seven hundred years to explain Stonehenge. It has been called variously a center for sun worship, a Druid temple, a Roman temple and a tomb. Excavation since 1920 has led to the belief that it was used for burial purposes and that it began in the New Stone Age. As we had lunch in one of the towns we passed, I noticed the sign "Cooperative" on some buildings and I also noticed a new three-wheeled automobile in a parking lot.

Bath, England is west of London and a resort area. In recent years they have discovered roman baths that had been buried under Bath since the Romans occupied the area. Excavation is continuing to uncover more of the hot springs that have been flowing for ages. At Glastonbury we visited the remains of a monastery built and managed by catholic monks. These remains are called Abbeys but actually they are the remains of large catholic churches destroyed during the reign of King Henry VIII.

On Tuesday, July 27, we drove to Stratford where we stayed in the Shakespeare Hotel. On our way to Stratford we stopped to visit the colleges in Oxford, England. The school in Oxford was also started by the catholic monks. Another stopping place was Glenhime Palace, the birthplace of Winston Churchill. Churchill did not want the pomp and ceremony so he was buried with his family in a very small church cemetery near Glenhime. We visited the place where Shakespeare was born, as well as Trinity Church. Tuesday evening we were treated to the play, "Much Ado About Nothing" in Shakespeare's Memorial Theatre. A most beautiful theatre and we enjoyed the evening very much. The next day of our tour took us to Chester, England where we stayed at the Queen Hotel. On this day we stopped at Coventry, England to visit the Coventry Cathedral. Coventry is an industrial city and on November 14, 1940, Coventry Cathedral died in the flames which destroyed the city around it. It was Hitler's attempt to destroy a city in one single operation from the air. The technique of destruction developed as the war proceeded until it reached its horrible climax in February 1945 when 135,000 people were burned to death in eight hours. Buildings were flattened or burning, streets were in chaos and the center of the ancient city had ceased to exist. Out of these arose the most beautiful cathedral I have ever visited costing about three million dollars. It appears to be part of an ecumenical movement involving numerous Christian faiths. Contributions to its construction came from service men all over the world.

Our tour guide was not familiar with the Queen Hotel in Chester and it was not a first class hotel. Chester is on the border with Wales. I was not aware that England has so many canals. While at Chester we toured north Wales along the Irish Sea. It was a beautiful park area. Saturday we checked out of the Queen Hotel and continued our tour north to York, England. We could not help but notice the amount of agriculture in England. The tour took us around Manchester and Liverpool, a very industrialized area. The tour guide reminded us that the weather man had been especially kind to us. York, like so many cities of Europe, had a wall around it. Congestion was noticeable in England. It was our understanding that cities observed a green belt around the city which the city could not annex. Our stay in York was at the Viking Hotel and it was strictly first class. On Sunday morning we attended mass at St. Wilfred Church not far from the hotel. Our group was taking a walking tour of the city but we decided to take a tour on our own later. We enjoyed visiting with our tour group and Irene kept us moving.

On Monday, August 2, our tour took us westward to a very popular resort called Bonness near Windermere Lake. We stayed at the nice Beech Hill Motel next to the lake.Next day we took a tour to Glasmere where Wordsworth was born and lived. We also took a cruise on Windermere Lake. This is a very beautiful part of England and attracts tourists and vacationers by the droves. On Wednesday our tour arrived at Largs, Scotland. On the way we stopped at the birthplace and home of Robert Burns, the great poet. Largs is on the west coast of Scotland. The hotel was old but strictly first class. We stayed in Largs for only one night and then proceeded to Edinburgh, Scotland where we checked in at the beautiful George Hotel. After dinner at the coffee shop we were treated to a welcoming party complete with pipes and highland dancers. We also witnessed the "Ceremony of the Haggis" for the first time. Our trip from Largs to Edinburgh was probably the most scenic drive we had on the entire tour. Fences around farms consist of stone walls all over the island. Scotland is mountainous and we visited "Loch Lomond" which means lake Lomond. Two of these lakes also provide the water supply for the City of Glasgow. The city of Edinburgh supplied the bus for our tours around this part of Scotland and Bob took his bus back to London. One chore Irene imposed on Bob on one of our tours in Scotland was when she asked him to stop the bus and pull some Scotland heather for our group. A tour of the city took us to another large castle on the top of a large hill that was a fortress originally. Saturday we took a tour south of Edinburgh. We saw a lot of flowers in southern Germany but the islands are just as beautiful. In addition to the flowers everything is so green, especially in Scotland. Our tour south of Edinburgh took us to the home of Sir Walter Scott, a palace. His library was huge and he also had a large collection of guns. We were surprised to learn, great granddaughters of Sir Walter Scott still reside there. We also visited the Adams Palace where we saw some irreplaceable paintings. Some relatives of the Adams reside there. We had an excellent buffet dinner in the coffee shop of our hotel before retiring. On Sunday morning August 7 we attended mass at St. Mary's Cathedral within walking distance of our hotel. The corner stone of the church indicated construction was completed in 1813. We checked out of our hotel at noon and took a plane to Dublin, Ireland. Before the tours started from London, I picked up a road map of England and one of Ireland. I could not help but notice that Northern Ireland was on both maps. Apparently the British Commonwealth consists of England, Wales, Northern Ireland and Scotland. Perhaps these areas are more like the Canadian provinces and exercise considerable control over their own affairs. In our tour of the islands we used American Express travelers checks and the exchange rate at the time was about $1.77 for one British Pound.

It took Air Lingus about an hour to get us to Dublin, Ireland, where we checked in at the Jurys Hotel for two nights. It is a rather large and very nice hotel. Sunday evening we were treated along with hundreds of other tourists at a welcoming banquet at the hotel followed by a program of Irish humor, song and dancing. The food was very good and the entertainment superb. On Monday we took a tour of Dublin visiting St. Patrick's Cathedral, which incidently is not a catholic cathedral, and visited the huge Trinity College Library. We saw the "Book of Kels." It was written and painted with outstanding expertise about 800 A.D. in a monastery scriptorium. Monday evening we attended a local play at the Abbey Theatre.

We left Dublin Tuesday and traveled to southwest Ireland through Kilarny and on to Hotel Dunloe Castle. This was a castle at one time, but has been acquired by American capital and made into an impressive hotel located in the center of spectacular landscaping. On Wednesday we circled the southwest penninsula of Ireland where we saw more of rural Ireland and a number of fishing villages. It was the first day that we encountered fog and rain for most of the day and visibility was poor. At some of our stops, fires had been lit in fire places and the fuel was peat. On this day we saw some peat fields. On Thursday, August 12, we traveled to Limerick and visited Blarney Castle on our way. I had a chance to kiss the famous Blarney Stone located at the top of the castle but decided to remain an observer. The Limerick Inn Hotel was at the edge of town and a very modern hotel. On our last night in Ireland we were treated to a medieval dinner at an old castle outside of Limerick. At dinner I sat next to two priests from Wisconsin. The costumes and the program reflected medieval times. It was enjoyed by all. Our bus took us to the Shannon Airport on Friday for our trip home on Air Lingus. The airport is a few miles from Limerick along the banks of River Shannon. We did a little shopping in the duty-free custom stores.

Based on conversations with acquaintances who served overseas in World War II, I expected England to have a much higher standard of living than Ireland. Our visit to Ireland found no such disparity. In comparing wages with the income tax of from 17% to 28% and a national sales tax (Value Added Tax) of about the same size, it would appear that American workers have a higher standard of living than either the workers of England or Ireland.

When we arrived at Kennedy on our return, a representative of Grand Circle was there to help us through customs and get us to our flights back home.

In summary, we left Kansas City Wednesday, July 21, 1982 and returned August 13, 1982. We spent eleven days in England, two days in Wales, four days in Scotland and six days in Ireland. A lot of the time was spent on sight seeing bus tours.

The buses appeared to be new and were equipped with tape players. Ireland has its own currency and at the time we were there $1.43 would buy one Irish pound. □

CHAPTER XII

My Philosophy

During the turbulent nineteen sixties, when university students streaked and burned college buildings, Dr. Murphy, President of Kansas University made this observation, "This is a feeling generation. Not a thinking generation." In other words, the behavior of that generation was dictated solely by emotions and not as a result of human reasoning. Animal behavior is dictated largely by emotions. It has been said, the human brain has a great deal to do with human emotions. Reasoning should be the basis of our convictions and will-power the instrument to apply those convictions to our emotions and behavior. This chapter reveals some of my convictions and suggestions for their implementation.

THE dictionary defines philosophy thus, "The study or science of the truths or principles underlying all knowledge. Wise composure throughout the vicissitudes of life." Omitting my basic philosophy or attitude toward the ramifications of life would deprive the reader of characteristics that undoubtedly contributed to my successes and my disappointments.

In our world, one does not live alone and how a person's life unfolds depends a great deal upon a person's relationship with the society in which he or she lives. There is no doubt that a person's success depends heavily upon being born with a good mind and a healthy body. Financial success and social success also improve with education and training. It has been my observation that none of these assure success unless a person possesses a good character. We were taught at an early age that the human being consists of mind and body. In our catechism it defined the human being as body and soul. The mind or the brain is the little computer through which it is possible to master ourselves and until we master ourselves, we really should not seek to master others. A person needs to have the courage to stand by one's convictions but not with a closed mind. One must be compassionate to those in need. A person of integrity will remain honest and truthful even if it hurts. To quote Douglas McArthur, "Remember the simplicity of true greatness, the open mind of true wisdom and the meekness of true strength."

It seems to me that belief in a Divine Being or an omnipotent power is not only my crede but appears to exist in the human being of every culture, country, island or tribe on this earth. My religion teaches that my tenure here on earth is temporary at best and that my eternal home is beyond. Recognizing that fact, tends to keep one from getting too attached to temporal things. It also helps us tolerate the monumental hardships that humankind imposes upon itself. It is hoped that one day we will have a Supreme Court that will recognize that prayer is private communication between humankind and its Diety and no power on this earth can prevail against it.

Failure of our society to recognize the importance of the family unit to our country is incomprehensible. The ABC network had a three hour special on changes in the family from two parents to single parent. A better definition might have been the destruction of the family unit. I think it is important to remember, there are two institutions that Communist Russia must dominate to continue in power. The family and the church. Are we traveling down the same road? At present our small children are the wards of babysitters, day-care centers and the streets. Tomorrow they may be the wards of the federal government as they are in Russia. Our elected representatives in Washington have found it convenient to ignore the family. When my wages were thirty-five cents per hour, the income tax exemption for a dependent child was $1,000.00. When inflation drove my wages up to $10.00 per hour, the income tax exemption for a dependent child was $600.00. Is there any wonder the family is being torn apart? Nothing will ever replace the programming of a baby's mind by a loving and caring father and mother. Allowing the family unit to disintegrate is in my opinion unforgivable and the cost to our country may be incalculable.

In discussing the success or failure of friend, schoolmates or acquaintances one often hears explanations such as, "Yes he was lucky. He did well financially." Another might say, "I was getting along fine but I got into the wrong crowd and they threw me off the track." "If my family had not been such a drag I could have made it," offers another. Granted there is an element of luck in anything we undertake. Granted further, success comes harder for some than others. Very often, however, the individual is simply trying to blame someone or something for his or her shortcoming. Sometimes the individual likes to feel sorry for himself or herself. It has been said, "Self pity is the greatest disease of humankind." Success comes most often to those who believe,"Where there is a will, there is a way." Or, "Don't leave for tomorrow what you can do today." Or, "A penny saved is a penny earned." Or, "Do unto others as you would have them do to you." Or, "Never say can't, American ends in I Can." I like the success explanation used by John Houseman in his ads, "They made it the old fashioned way, they earned it."

Humankind resides on a few patches of land imbedded in a liquid marble that rotates once in every twenty-four hours in relation to the sun. For some unknown reason the magnetic poles are over land and water one hundred and eighty degrees apart. Scientists are still trying to find answers not only to the marble but to the extent of the universe. Our relationship with people of different backgrounds and different governments becomes our foreign policy. Under the present administration for example, we encourage free trade among nations. Some question how free that trade is. Besides the trade problem, we have been in a tug-of-war exporting our democratic form of government to other nations while Russia is exporting its socialistic form of government to other nations. While Russia calls itself socialist, it is just another form of dictatorship. Ten percent of the people dominate ninety percent of the people. Democracy works fine in a country where the voter can read and write but not very well where the society remains illiterate. We must trade and deal with countries as they are and not what we would like them to be. In dealing with Russia, we should remember what former President Truman said, "We had twenty-six agreements with Russia after World War II and they broke every damn one." I hope our government leaders watch what Russia does and react accordingly. In college we debated the issue of international free trade. That theory contends for example, if the United States can produce wheat cheaper than any country, the remainder of the world should buy our wheat. On the other hand, if the Arab world can produce oil cheaper than any country in the world all countries should buy oil from the Arabs. Do you remember what happened when we bought most of our oil from the Arab states? There probably never was and never will be such a system of international free trade. It will always be a matter of nations negotiating with each other. If we had international free trade, no duties, no subsidies, would our standard of living be reduced to a world level?

We have had two political parties, Republican and Democrat, all of my life and I have never taken an active role in either party. I am not enamoured with either party. Like Eisenhower, when both parties sought him for their presidential candidate, I think each party's position on important matters is extreme. Ike said, "The Republicans drive in the right ditch, the Democrats drive in the left ditch and that's where the gutter is." The democrats would be happy if the federal government owned and controlled everything. The republicans would be happy if the federal government owned and controlled nothing. What has emerged from such an arrangement is that neither party elects a president. People like myself have to cast our vote for a republican or a democrat who we hope will modify the party extremes. It is a shame that the people of this great country cannot agree on the proper role of the federal government and private enterprise. There is no doubt in my mind that private enterprise, with its profit motive, is desirable but it must be tempered by a federal government that acts like an umpire. Neither a government monopoly nor a private monopoly must be allowed to develop. Just as I favor private competitive enterprise, I also favor turning many of our programs over to the states when they are closer to the problems. Instead of mailing our income tax to Washington and letting congress decide how much to return, why not let us in Missouri handle the problem with our own income tax and let us deduct our Missouri income tax from the amount of federal income tax due. The federal income tax would be the maximum income tax we pay to be apportioned between federal, state and local entities. Government at all levels is administrative in nature with few exceptions. It produces few products or services on which our standard of living depends. When government absorbs too much of our resources the economy suffers. Government has grown by leaps and bounds over the last fifty years.

The disturbing part about private enterprise is that while it offers incentives for individuals to work, save and invest, private enterprise has a tendency to ignore other obligations beyond making a profit. For example, private enterprise will discharge thousands of employees, to increase its profits, but feels no obligation to find a job for those people. Private enterprise does not hesitate to reduce wages of its employees, to increase profits, but shows very little compassion for the affect it will have on the standard of living of those employees. The federal government must have the authority to own and control certain assets and resources, to provide necessary public improvements and to see that private enterprise does not abuse the rights of people. The government of the United States does not owe me a living but I contend it owes me an opportunity to make a living, if I am willing and able to work.

Part of the enormous growth in government at all levels is due to the eagerness of our elected officials to retain their jobs.

One way to reward people who worked diligently in one's election campaign is to get them a job on a committee or an agency created by congress. After that has been accomplished you must work hard to keep those agencies going. As a result we have been adding committees or agencies and seldom do we eliminate an agency. A certain percent of our population will always be unemployable due to mental or physical handicaps and those of us more fortunate must provide for those people through federal and state agencies. This constitutes a sizeable voting constituency and when a federal agency inflates this number by adding to the welfare rolls many who are not disabled, it becomes a vote-buying process and not a welfare program. Individual initiative is destroyed, when it pays to be unemployed.

The depression years in the thirties brought many changes in our way of life and most of it started by bestowing unusual powers on the federal government. The social security system was enacted to provide for the elderly as other civilized nations had done. Unfortunately, that system has become a political football. We were told by our elected officials, a fund was being created by contributions from employees and employers to provide retirement benefits. There has never been such a fund. More and more groups were added to the system without compensating contributions to the fund. Benefits have been paid to applicants completely out of proportion to the premiums received from the applicant. Current collections from the system were used by our elected representatives to create a thirty-five million voter constituency of elderly people. It has become a monstrous vote-buying process. The benefits are neither pension benefits nor welfare benefits. Contributions to the fund, plus incremental earnings on the fund, fall far short of meeting the benefit obligations. If it is a retirement pension benefit, the payment should be related to the premiums paid for the benefit. Since that is not the case, the payment is really a welfare benefit to the elderly without any evaluation of need. Millionaires are receiving social security benefit payments and under present income tax statutes pay income tax on only half of the benefits. Since I qualify for social security benefit payments and have paid the maximum premiums in the system since enactment of the law, how would I have fared had I purchased an annuity policy with the money and the interest rate had been six per cent? According to my calculations, my contributions plus the contributions of my employers, invested at six percent compound interest would have financed my social security benefits for a period of a little over three years. Is there any wonder the system is bankrupt?

Congressional elections are being held this year and a presidential election will be held in two years. The Reagan Administration will no doubt take full credit for the decrease in interest rates, decrease in inflation and the decrease in unemployment since it came into power. Neither party wants to talk about the budget deficit, the international trade deficit and the unbelievable trillion dollars of national debt accumulated during the present administration. I have read Milton Friedman's nobel prize winning "Free to Choose" and referred to some chapters many times. I also read the Wall Street Journal daily. Economic conditions always go into high gear when billions of dollars are thrown into the economy such as in war. Is it possible that we are having a real fling today and the inevitable hangover will appear tomorrow? The achievements mentioned above are real but we may have to wait for the eventual costs of those achievements. No solution has been found for our budget deficit, no solution has been discovered for our international trade deficit, no solution has been found for our bankrupt agriculture and no attempt has been made to stop the mushroom growth in our trillions of dollars of national debt. The latter is especially troublesome when ten percent of it is payable to one foreign country, Japan.

During my many years with cooperative organizations, a representative from the Farm Credit Administration once posed the question, "Just how democratic are our cooperatives?" This had reference to the apparent lack of interest the members had in their cooperative. Many annual meetings of members would not attract enough members to constitute a quorum to transact business. The same question might be asked today, "Just how democratic is this system that we want to export to other nations?" Do our elected representatives become statesman and vote for their country or do they surrender their vote to special money interests? Since individuals are limited in the amount one can contribute to a favorite candidate, we invented the Political Action Committee.

Every special interest group, including director and employees of REA Borrowers, have their own PAC Committee through which millions of dollars are funneled to the candidate who has voted consistently for federal gratuities to the particular special interest group. Business, labor, education, doctors, lawyers, CPAs and hundreds of others are rewarding a congressman or senator by large contributions to his or her political campaign. Those with large campaign funds and plenty TV exposure generally get elected. Last week the press quoted a California congressman as saying, "The system at times forces them (Members of Congress) to do things that they would rather not do." He is quoted further, "I think that process buys you out. But I don't think that you individually have been bought out, or that you sell out. I think there's a big difference there . . . One is morally wrong, in effect. The other is legally wrong. There's a big difference." A senator from

Arizona is quoted as complaining, "It is not 'we, the people' but political-action committees and moneyed interests who are setting the nation's political agenda and are influencing the position of candidates on the important issues of the day." A senator from Mississippi concurs, "We are gradually moving elections away from people, as certainly as night follows day." It appears that "We, the People" have been electing politicians instead of statesmen to Washington. If the situation is that bad, our congressmen and senators can become statesmen by passing legislation to cut off the funding.

Another problem has made itself known in the last few years and is wrecking insurance protection for everybody. Insurance has for many years been available for life, property and liability claims. By many people paying premiums into one insurance fund, claims can be paid to the unfortunate for loss of life, property or for liability damages. In recent years liability claims have been taken to courts and millions upon millions of unwarranted settlements awarded to plaintiffs. Insurance rates have multiplied many times and insurance coverage has been denied in many cases. I know the trial lawyers' association wants to blame industry, doctors, CPAs or the insurance companies. It is my understanding that the United States is the only major nation in the world that allows lawyers to share in the liability awards. Some lawyers become millionaires by winning one liability case. In some cases the lawyer gets most of the proceeds. The solution is simple. Cut the lawyer out of the awards.

This brings to mind a conclusion reached by a lawyer regarding his profession. I rented my first office space from a real estate broker by the name of George Crum, who earned a degree in law but never practiced law for a livelihood. Both of us were in the office when George was reaching exasperation in a real estate transaction involving another lawyer. In frustration he raised his voice and asked "Walt, do you know what's wrong with lawyers?" After weighing my answer I asked, "Since you are a lawyer, what is wrong with lawyers?" "They're crooked" George remarked.

The Declaration of Independence contends that the Creator endowed the human being with the unalienable right to life, liberty and the pursuit of happiness. The world recognizes the progress we have made in the protection of human life and the safeguarding of our individual freedom. In one of his editorial writings, Bishop Fulton J. Sheen questioned whether happiness was actually the answer to the yearning of humankind. He suggested that peace of mind was the real objective. Using a word from the Declaration of Independence, it is "self-evident" that the scrambling for wealth, power or fame fails miserably to result in providing a life of contentment and peace. In another editorial article Bishop Sheen made this observation, "contentment begins with self-abandonment." Self-pity and depression is frequently a matter of mental attitude. Losing oneself in an enjoyable occupation, dedication to providing for those dependent upon us and leading a life pleasing to God and to our fellow-man, are the road signs to contentment and peace.

A favorite prayer of mine, "Lord grant me the serenity to accept what I cannot change, the courage to change what I can and the wisdom to know the difference."

BIBLIOGRAPHY

MIGRATIONS:

THE GOLDEN JUBILEE OF THE GERMAN RUSSIAN SETTLEMENTS OF ELLIS AND RUSH COUNTIES, KANSAS compiled by B.M. Dreiling of Hays, Kansas and published by the Ellis County News.

THE CONQUERING WIND, written by Amy Brungardt Toepfer and Agnes Dreiling of Ellis County, Kansas, published by Victor C. Leiker.

DIE LIEBENTHALER UND IHRE KIRCHE a history of Liebenthal, Kansas and St. Joseph's Catholic Church, written by Emerald Dechant.

RURAL ELECTRIFICATION:

THE FARMER TAKES A HAND written by Marquis Childs and published by Doubleday & Company, Inc. in 1952.

KANSAS RURAL ELECTRIC COOPERATIVES—TWENTY YEARS WITH REA compiled by The University of Kansas in 1960.

CPA PROFESSION:

Various publications of the American Institute of Certified Public Accountants.

THE BIG EIGHT written by Mark Stevens and published by Macmillan Publishing Co. Inc. in 1981.